U0955526

Bereavement Guilt in Chinese Adult Children: The Preliminary Exploration

Li Jie

知识产权出版社
全国百佳图书出版单位

图书在版编目(CIP)数据

中国丧亲者的内疚感：当父母去世 = Bereavement Guilt in Chinese Adult Children：The Preliminary Exploration：英文/李洁著. —北京：知识产权出版社，2016.7

ISBN 978-7-5130-4312-0

Ⅰ.①中… Ⅱ.①李… Ⅲ.①成人心理学—研究—中国—英文 Ⅳ.①B844.3

中国版本图书馆 CIP 数据核字(2016)第161829号

责任编辑：常玉轩　　**责任校对**：谷　洋

装帧设计：陶建胜　　**责任出版**：孙婷婷

Bereavement Guilt in Chinese Adult Children：

The Preliminary Exploration

李　洁　著

出版发行：知识产权出版社有限责任公司	**网　　址**：http://www.ipph.cn
社　　址：北京市海淀区西外太平庄55号	**邮　　编**：100081
责编电话：010-82000860 转 8572	**责编邮箱**：changyuxuan08@163.com
发行电话：010-82000860 转 8101/8102	**发行传真**：010-82000893/82005070
印　　刷：北京中献拓方科技发展有限公司	**经　　销**：各大网上书店、新华书店及相关专业书店
开　　本：720mm×1000mm　1/16	**印　　张**：12.75
版　　次：2016年7月第1版	**印　　次**：2016年7月第1次印刷
字　　数：130千字	**定　　价**：45.00元

ISBN 978-7-5130-4312-0

本成果受到中国人民大学“统筹推进世界一流大学和一流学科建设”专项经费的支持。

谨以本书献给我的母亲张丽荣女士

Abstract

Bereavement is one of the most painful experiences in one's life, and guilt is one of the various emotions associated with it. Guilt is widely reported by bereaved people, but it is not yet sufficiently understood by researchers. There is a lack of convergence in conceptualizing guilt, and no valid measurement exists to assess this important construct in grief. Meanwhile, this emotion is particularly significant in Chinese adult given the cultural value of filial piety. Therefore, efforts are needed to fill this gap and enhance understanding of the nature and consequences of bereavement guilt in Chinese adult children.

Several studies are included in this book. This sequential mixed method research comprises three parts. To better define and conceptualize bereavement guilt, 16 bereaved adult children were interviewed in the first study about their subjective experience of guilt. Their narratives illustrated the content and feelings of bereavement guilt. It suggested that guilt was based on various evaluations of their behaviors, which failed to meet their inner standards or expectations. It also revealed that they felt guilty for contributing to the death, for hurting the deceased, and for not paying enough back to the parents. Those thoughts were evidently accompanied by distressful feelings.

In the second study, the Bereavement Guilt Scale (BGS) was developed based on findings in study one. It was then validated with survey data collected from 873 bereaved adult children. The scale comprises five factors: (1) responsibility towards the death; (2) hurting the deceased; (3) survivor guilt; (4) indebtedness guilt;and(5) guilt feelings. It was proved to have good dimensionality and construct validity, satisfactory internal consistency and test – retest reliability.

In the third study, more explorations on bereavement guilt have been conducted, including its prevalence, predictors and relationship with complicated grief. The same data set in study two was analyzed and the results revealed various occurrence rates for different subtypes of bereavement guilt. The most common type is indebtedness guilt(81%), followed by guilt feelings(70%), responsibility for the death(37%), survivor guilt(27%) and guilt about hurting the deceased (23%). It was also revealed that guilt is more commonly experienced by people bereaved from unnatural death than natural death. Investigating a group of demographic and death – related variables simultaneously, this study found that different types of guilt have distinct predictors, but that three risk factors predicted most of them. These are the adult children being of a younger age, losing a mother, and low psychological preparation for the death. Finally, a strong association between guilt and complicated grief was established by the data. The relationship between complicated grief and guilt was stronger than with demographic variables, death – related variables, depression and anxiety. Such association remains significant even when controlling for these variables.

In summary, the author proposed a multidimensional conceptualization of bereavement guilt and developed a valid measurement tool. It built a preliminary model of risk factors, bereavement guilt and health outcomes in bereavement. The implications and limitations are also discussed.

Content

Chapter 1 Introduction 1

Bereaved adult children: The gap in literature 3

Guilt: An important emotion in grief 4

The present studies: Guilt in bereaved Chinese adult children 6

Overview of the project 8

Chapter 2 Literature review 11

Guilt in bereavement 13

Critical analysis and research gaps 27

The proposed conceptualization of bereavement guilt 35

Chinese Bereaved Adult Children 37

The present research 41

Chapter 3 Methodology 45

Meeting research objectives 47

What is mixed method research 49

Advantages and functions of mixed method 50

Paradigm debate and pragmatism 51

Methodological Framework 53

Chapter 4 The subjective experience of guilt: Insiders' voices 55

Introduction 57

Method 59

Results ······ 66

Discussion ······ 79

Chapter 5 Development and validation of a Bereavement Guilt Scale ··· 85

Introduction ······ 87

Method ······ 88

Results ······ 95

Discussion ······ 114

Chapter 6 Bereavement guilt: Its prevalence, predictors and relationship to complicated grief ······ 119

Introduction ······ 121

Method ······ 124

Results ······ 125

Discussion ······ 138

Chapter 7 General discussion and conclusion ······ 151

Summary of the research and findings ······ 153

Contributions and implications ······ 156

Limitations ······ 166

Future directions ······ 168

Reference ······ 171

List of Tables

Table 2. 1 Quantitative Studies of Guilt in Bereavement ··············· 16

Table 3. 1 A Pragmatic Alternative to Key Issues in Social Science Research Methodology ·· 52

Table 3. 2 Methodological Framework ······································ 54

Table 4. 1 Characteristics Variation in Sampling ························ 60

Table 4. 2 Background Information of Participants ······················ 62

Table 5. 1 Demographic and Death Related Characteristics ············ 93

Table 5. 2 KMO and Bartlett's Test ··· 96

Table 5. 3 Total Variance Explained by Initial Four Factors ············ 97

Table 5. 4 Communalities of Initial EFA ··································· 97

Table 5. 5 Rotated Factor Loadings for Initial 26 Items of Guilt Cognitions ··· 99

Table 5. 6 KMO and Bartlett's Test for Final EFA ····················· 100

Table 5. 7 Total Variance Explained by Four Factors in Final EFA ··· 100

Table 5. 8 Communalities for Final EFA ·································· 101

Table 5. 9 Rotated Factor Loadings in Final EFA ······················ 102

Table 5. 10 Item – Total Correlation ······································· 103

Table 5. 11 Inter – Item Correlations ······································ 104

Table 5. 12 Inter – Item Correlation of Guilt Feeling Items ··········· 105

Table 5. 13 Item – Total Correlation of Guilt Feeling Items ··········· 106

Table 5. 14 Model Fit Indices for Three Competing Models ··········· 111

Table 5.15 Correlations Among Factors in Model 2 ········ 112
Table 5.16 Correlations of the BGS with Self – Esteem and Self – Forgiveness ········ 112
Table 5.17 Correlations of BGS with Depression and Anxiety ········ 113
Table 5.18 Internal consistency and test – retest reliability of BGS ··· 114
Table 6.1 Prevalence of Guilt in Total Sample ········ 126
Table 6.2 Prevalence of Guilt in Different Subgroups ········ 127
Table 6.3 Standardized Coefficients of Predictors of Guilt in Simple Regression Analysis ········ 128
Table 6.4 Standardized Coefficients of Predictors of Guilt in Multivariate Regression Analysis ········ 129
Table 6.5 Hierarchical Regression Analysis of Complicated Grief on Background Variables and Bereavement Guilt ········ 132
Table 6.6 Hierarchical Regression Analysis of Complicated Grief on Background Variables, Psychological Distress and Bereavement Guilt ········ 133
Table 6.7 Correlations Between Types of Bereavement Guilt and Complicated Grief ········ 134
Table 6.8 Regression Coefficient of Each Guilt Component Controlling for Background and Psychological Distress Variables ········ 135
Table 6.9 Multiple Regression Analysis of Bereavement Guilt to Complicated Grief ········ 136
Table 6.10 Regression Analysis to Test the Mediating Effect of Guilt Feelings ········ 137

List of Figures

Figure 2.1 Research Framework ············ 42
Figure 4.1 Aspects in Phenomenological Experience of Bereavement Guilt ············ 67
Figure 4.2 Themes of Guilt Experience in Bereavement ············ 68
Figure 5.1 Scree Plot of Initial EFA ············ 97
Figure 5.2 Scree Plot for Final EFA ············ 101
Figure 5.3 Theoretical Structure of Bereavement Guilt Model 1 ······ 107
Figure 5.4 Theoretical Structure of Bereavement Guilt Model 2 ······ 108
Figure 5.5 Theoretical Structure of Bereavement Guilt Model 3 ······ 109
Figure 7.1 Risk factors, Bereavement Guilt and Health Outcomes ······ 159
Figure 7.2 Structural Model for Investigating Bereavement Guilt ··· 168

Chapter 1

Introduction

"*Guilt is the hilt of the knife that we use on ourselves, and love is often the blade*"

G. D. Roberts(2003): Shantaram(p. 426)

Bereaved adult children: The gap in literature

Losing a loved one through death is one of the most distressful events that can occur in life, and one which most people will unfortunately experience during their own lifetime. In China, nearly 10 million people die each year, leaving an enormous number of people in bereavement[1]. With the improvement of medical and health conditions, people are enjoying a longer life today than they did decades ago. The life expectancy of Chinese people is now more than 70 years. Since most people die at the old age, most people lose their parents when they are adults. Therefore, bereaved adult children may make up a large proportion of the grieving population.

However, bereaved adult children have not attracted enough attention from researchers and clinical professionals. There are relatively fewer studies focusing on this group of bereaved people, compared with other subgroups. For example, in the latest handbook of bereavement research (M. S. Stroebe, Hansson, Schut & Stroebe, 2008), there are specific chapters about bereaved spouses, young bereaved children, bereaved parents, even the bereavement of grandparents, but not about bereaved adult children. A similar situation is evident in other recent

[1] Bereavement refers to the objective situation of having lost someone significant.

influential publications in the area of bereavement (Robert, Neimeyer, Harris, Winokuer & Thornton,2011).

This situation is probably due to the impression that losing a parents in one's adulthood is a natural process, thus bereaved adult children are supposed to manage well. However, given the significance of the parents - children relationship and the challenging nature of bereavement, losing a parents even in one's adulthood can also become an extremely stressful and challenging event and impose considerable influence on one's well - being (Umberson & Chen, 1994). In order to fulfill this important gap, the present thesis will focus on the experiences of bereaved adult children.

Guilt: An important emotion in grief

Grief refers to the diverse reactions contained within the process of bereavement. There are different emotional manifestations in grief, including depression, despair, dejection, anxiety, fear, guilt, anger, hostility, loneliness, yearning, longing, shock, numbness and so on (M. S. Stroebe, Hansson, Stroebe & Schut, 2001). The occurrence, and intensity of those emotions will differ across individuals, bereavement time and even different cultures. In the recent two decades, the scientific research on bereavement has rapidly developed, and more detailed understandings regarding grief reactions and consequences have been accumulated. Researchers are beginning to explore understanding related to different emotions contained in grief; such as loneliness (J. Costello, 1999; Stewart, Craig, MacPherson & Alexander, 2001; W. Stroebe, Stroebe, Abakoumkin & Schut, 1996) and anger (Drenovsky, 1994; Exline, Park, Smyth & Carey, 2011; S. H. Kim, 2009). Moreover, based on a series of empirical findings, some emotions occurring in separation distress, such as yearning and loneliness; and in traumatic distress, such as anger and bitterness, are supposed to be manifestations of complicated grief(Prigerson et al. , 1999)[1]. Even though so far some emotions in grief have been investigated more often than others, the developing trend of this field will eventually bring a sophisticated understanding and knowledge of each significant emotion in grief. Among the various types of emotion in grief, guilt is a significant and special one which deserves a more systematic understanding.

Guilt is a long recognized emotion in bereavement. Even in the famous article *Mourning and Melancholia* by Freud(1917/1963), guilt was mentioned as a reaction for those who have lost loved ones. It has been acknowledged by bereavement researchers as early as the 1940's(Lindemann, 1944), and also in a clinical case report on professional helpers who worked with bereaved clients(Tooley, 1975). It is a familiar emotion within grief, with a long history of documentation.

The significance of guilt is also demonstrated from its correlation with psychopathology. Excessive and exaggerated guilt is one of the diagnostic criteria of Major Depression Disorder(APA, 1994). As will be seen in the literature review, a considerable number of studies have established a close relationship between guilt and psychological disorders; such as depression, anxiety disorder, or traumatic reactions, such as PTSD.

Unlike other emotions(e. g. sadness) in grief, guilt is a self – conscious emotion (Tangney & Fischer, 1995). It is based on the self reflection and self – evaluation of oneself and one's own behavior(Leary, 2007). The feeling of guilt is generated from the perceived gap between inner standards and actual behaviors, perceiving that one has done something wrong or has failed in something one was supposed to do. Therefore, guilt bears with it a judgemental nature, and is usually accompanied by self

[1] Complicated grief is a pathological syndrome of grief reactions, which will be introduced in latter section.

reproach and self blame. This self attack is a threat to one's self – esteem, and may also add extra stress and risk to a bereaved person's well – being.

In the context of losing a significant loved one through death, guilt is a relevant emotion because it also has an interpersonal nature (Barrett, 1995). It is an emotion that is not only connected to intrapersonal processes, such as self – evaluation, but also portrays interpersonal functions. Psychologists have argued that the function of guilt is to preserve and enhance a relationship. Guilt prompts people to avoid transgressions against the relationship. It also motivates people to make amends to the other person in the relationship, and to redistribute the distresses that arise in the relationship, minimizing inequality (Baumeister, Stillwell & Heatherton, 1994). It is postulated then that guilt may play an important role in adjustment within bereavement, which implies the loss of both a significant other and the relationship itself.

The present studies: Guilt in bereaved Chinese adult children

Apart from the above evidences and considerations, studying guilt in bereaved Chinese adult children is particular meaningful, given the importance of the emotion guilt and parents – child relationship in Chinese culture.

In a study which investigates the organization of emotion across western and eastern countries, guilt and shame are placed as basic emotions only among Chinese participants, not among participants from other countries (Shaver, Wu & Schwartz, 1992). This indicates that guilt is a more familiar and significant emotion in Chinese culture than in western culture. Meanwhile, guilt is described in a more sophisticated way in Chinese than in English, as the different Chinese words describing guilt do not have direct corresponding translations in English(Bedford & Hwang, 2003).

In collectivist societies, such as China, interrelatedness is emphasized. The identity of an individual is defined by relationships and embedded in roles (U. Kim, 1994). The relationships within the family and the roles of each family member form the foundation of the social order. Among those relationships, the parents – child relationship is one of the most important. For example, the three cardinal guides (ruler guides subject, father guides son and husband guides wife) in traditional Chinese culture, exemplify the close bond in the parents – child system. Filial piety, another important cultural value, prescribes a series of strict and concrete role obligations of children to their parents. There is no direct translation of grief in Chinese, but there is a close term: *shou xiao*, which literately means expressing filial piety (Klass & Chow, 2011). According to traditional customs, a bereaved adult child should *shou xiao* for 27 months, symbolizing paying his/her parents back for their care and nurture during the first 27 months of the person's life. During that period of time, bereaved adult children should live in a self – restricted way. They are prohibited from dancing, singing, marrying and even eating meat.

In conclusion, the relationship between parents and child is highly emphasized in Chinese society. In Chinese culture, values and rules exist to regulate the obligations and duties of adult children to their parents, either before or after the parents' death.

Given the great number of bereaved adult children, the significance of guilt and the parents – child relationship in Chinese culture, and the limited understanding demonstrated in existing literature, investigating guilt in bereaved adult children is a particular meaningful attempt to understand the bereavement and grief experience for Chinese people.

Overview of the project

Even though guilt is a common and significant emotion in bereavement, the understanding of its nature and consequences is fragmented. This project aims to contribute more knowledge toward this topic, in order to broaden the knowledge of the grief experience and to provide inspiration for helping bereaved people in need.

Previous studies on guilt in bereavement mainly involve three main themes: the prevalence of guilt, the risk factors associated with guilt and the relationship between guilt and health outcomes in bereavement. However, as will be analyzed in Chapter 2, problems and shortcomings in conceptualizing and measuring guilt are evident in those studies, and they have contributed to the discrepancies and limited findings regarding these research questions. There exists no appropriate conceptualization and scale for bereavement guilt that can be borrowed from elsewhere to this field, because there is generally a lack of convergence in psychology on defining and measuring guilt. With the research trend and necessity of developing a domain specific understanding of guilt, it is timely to develop an appropriate conceptualization and measure of bereavement guilt. In this project, a multidimensional conceptualization of bereavement guilt will be proposed, and a valid scale for bereavement guilt will be developed and validated.

With this new understanding and tool related to bereavement guilt, more understanding on the nature and consequences of bereavement guilt will be explored. First, statistical results will be provided on how common guilt exists in bereaved Chinese adult children. Second, a group of risk factors will be examined in order to profile which group of people is at risk. Last but not least, the relationship between guilt and complicated grief will be explored.

Complicated grief is an emotional syndrome which is considered to be a pathological grief reaction. Not every bereaved person will develop complicated grief, and the prevalence ranges from 7% to over 40% across difference subgroups of people (Kersting, Brähler, Glaesmer & Wagner, 2011; Newson, Boelen, Hek, Hofman & Tiemeier, 2011; Ott, 2003; Prigerson et al., 1995). Efforts have been made to propose the inclusion of complicated grief (or as known by another name, prolonged grief) as a distinct disorder in the DSM – 5 (Prigerson, Vanderwerker & Maciejewski, 2008). Even though complicated grief is correlated with the symptoms of depression and anxiety, it is proved to be a distinct construct (Bonanno et al., 2007; Prigerson & Frank, 1995). Complicated grief is associated with prolonged distress, both psychological and physical problems, and suicidality (for a review, see M. K. Shear et al., 2011). Exploring correlates of complicated grief has been a central topic in recent years (e. g. Kristensen, Weisæth & Heir, 2010; Lobb et al., 2010; Melhem et al., 2004), and increasing effort has been made to explore and discuss its development process (Boelen, van den Hout & van den Bout, 2006; K. Shear et al., 2007). Examining the relationship between guilt and complicated grief will certainly contribute to this pool of knowledge.

The literature in Chapter 2 will summarize previous findings, identify gaps and introduce the research objectives and research framework of the project.

Chapter 3 explains the methodology of the project. In order to meet research objectives and answer each research question, this project adopts a mixed study design. The rationale and philosophy behind adopting mixed methods will be introduced, along with the methodological framework.

Chapter 4 describes the qualitative study, which aims to explore the subjective experience of guilt in bereaved adult children, and collect narratives for defining bereavement guilt and developing a Bereavement Guilt Scale (BGS).

Chapter 5 includes the development and validation of the BGS, assessing its psychometric properties and confirming the structure of bereavement guilt.

Chapter 6 applies the BGS to construct empirical understandings about guilt in bereaved Chinese adult children, including the following three aspects: its prevalence, predictors and relationship to complicated grief.

The research objectives, methods, results and discussions of each study will be presented in chapter 4, 5 and 6 respectively. Finally, chapter 7 presents conclusions summarizing the findings of each chapter, along with an integrative discussion of the whole project.

Chapter 2

Literature review

This chapter will review existing studies on guilt in bereavement. A critical analysis on those studies will reveal gaps and limitations of previous research, involving definition ambiguity and measuring difficulties that prevent further understanding. It will then illustrate why a specific conceptualization of guilt in bereavement is necessary based on summarizing existed conceptualizations and measures of guilt in general psychology. Based on the review, a definition of bereavement guilt will be proposed, and the research framework of the project will be introduced in the end.

Guilt in bereavement

Long Recognized Phenomenon

The reactions to loss are various, and include different aspects. Grief is understood by many researchers to be an emotional syndrome, incorporating diverse psychological and physical reactions (e. g. M. S. Stroebe et al., 2001). Guilt is a familiar emotion within the grief syndrome, long – recognized by writers and widely reported by bereaved persons themselves. As early as the 1940s, guilt was included as one of the main grief reactions, one in which people search through their memories for wrongdoings and omissions to the deceased (Lindemann, 1944). Guilt – related feelings have also been identified in qualitative studies (observations, in – depth interviews and focus groups) as one of the main themes in people's grief experience (Adolfsson & Larsson, 2004; Cacciatore, 2010; Clayton, Halikes & Maurice, 1971; Kerr, 1994; Miles & Demi, 1983, 1991, 1992). For example, Smialek (1978) noted that self – reproach and guilt are among the most common reactions of bereaved parents who have recently lost an infant suddenly or unexpectedly. Self – blame for not doing enough and survivor guilt were also main themes emerging from a focus group study among bereaved elderly persons (Smith, Nunley, Kerr & Galligan, 2011). Furthermore, guilt has been considered an important aspect of grief both as a normal reaction(Shuchter & Zisook, 1993) and as a manifestation of pathological grief(Parkes, 1996; Raphael, 1975), even being considered a reason for the development of complications in grieving(e. g. , Rando, 1993). The centrality of this emotion is also reflected in its incorporation as a subscale in classic grief measures, such as the Grief Experience Inventory(GEI) (Sanders, Mauger & Strong, 1985) and the Grief Experience Questionnaire(GEQ) (Barrett & Scott, 1989). Thus, there are good reasons to argue the need for sound empirical investigation of the phenomena and manifestations of guilt in bereavement.

Existing Empirical Findings

What empirical findings of guilt have been accumulated in the domain of bereavement? The literature search identified 34 relevant published quantitative studies. These studies are summarized in Table 2. 1. The studies about guilt in bereavement can be summarized into three domains: the prevalence of guilt, risk factors of guilt, and the relationship between guilt and health outcomes in bereavement.

The Prevalence of Guilt

Guilt has been reported as one of the main themes mentioned by bereaved people in a lot of qualitative studies, but how common is it according to quantitative studies? The prevalence reported ranges enormously, from 7% to over 90%, but most are around 30% ~ 60%. There are two possible reasons accounting for this diversity. The first has to do with the nature of the particular subgroups under investigation. The two studies reporting over 80% for guilt were conducted among samples of bereaved people where there is likely to be particular vulnerability to guilt feelings. One was among parents of hemophiliac children (Mizota, Ozawa, Yamazaki & Inoue, 2006). High levels of guilt among these parents may have to do with the fact that most of the children inherit the disease from their mothers. The second study involved parents in support groups (Miles & Demi, 1991, 1992). It would seem plausible that parents seeking support would do so for help in dealing with difficult emotions such as guilt feelings, such that the prevalence would be higher than for bereaved persons in general. In contrast to these studies, the very low percentage (7%) was reported for spouses bereaved for a longer time (above 4 years) (Zisook & Shuchter, 1985).

Second, measurements of guilt vary remarkably between studies. The high prevalence ones typically count various sources of guilt (such as not doing enough, responsibility for the death, survivor guilt and so on) rather than one specific aspect. This naturally increases the chance of guilt being detected. For example, the article with the low percentage, by Zisook and Shuchter (1985), did not give a clear definition of guilt, and left the interpretation open to participants. Moreover, some guilt questions are in yes/no format (Davis, Lehman, Wortman & Silver, 1995; Jonasson et al., 2011; Ylitalo, Valdimarsdottir, Onelov, Dickman & Steineck, 2008), while others ask about the frequency of guilt feelings (Surkan et al., 2006). Thus, the prevalence of guilt bears different meanings in different studies, and it is difficult to make comparisons across them.

In conclusion, what can be said is that most studies show guilt to be a common phenomenon among bereaved people. However, the results indicate that no general conclusion about prevalence of guilt in bereavement is possible or probably even useful. What we need is better understanding of prevalence for specific subgroups, taking risk factors (see below) into consideration. Also, we need to establish precisely what is meant by guilt and how to measure it accordingly.

Table 2.1 Quantitative Studies of Guilt in Bereavement

Authors/Location	Method	Main Measurements	Main Findings about Guilt
Shanfield and Swain (1984)/US	Xs./40 parents, lost children in car accident 13 – 36 months ago	Guilt: One item about feeling of guilt; H. O.: SCL-90, BDI, Self – constructed grief scale	Prev. 42.5%; R. F.: Ambivalent relationship before death(+ ve assoc.)
Horowitz et al. (1984)/US	Long. (T1: 2M, T2: 7M, T3: 13M)/72 adult children(35 clinical sample & 37 filed sample)	Guilt: One item on self – blame for the death; H. O.; IES, Clinician's ratings of stress – specific signs and symptoms, SCL – 90, ESS	Guilt & decrease of psychological distress: – ve assoc.
Zisook and Shuchter (1985)/US	Xs./300 spouses; 47% lost someone from prolonged illness	Guilt: One item on self – reproach about the death; H. O.: Self – constructed grief scale	Prev.: 7% (bereaved more than 4Y) to 26% (bereaved less than 1Y); R. F.: Time(N. S.)
Lehman and Wortman (1987)/US	Xs./37 spouse & 41 parents; lost family members in motor crash	Guilt: One item about perceived preventability	Pre: Ever since the death: 52% (spouses) & 62% (parents); In the past month: 28% (spouses) & 28% (parents)
Downey, Silver, and Wortman(1990)/US	Long. (T1: 1M, T2: 3M, T3: 18 M)/228 parents who lost children in SIDS	Guilt: One item on self – attribution for the death; H. O.; SCL – 90 – R	Guilt & psychological distress: + ve assoc. (concurrent); N. S. (longitudinal)
Bohannon(1991a)/US	Long. (T1: 2M – 5Y, T2: 6 M after T1, T3: 12 M after T1)/33 couples who lost their children	Guilt: Sub – scale in GEI	R. F.: Time (– ve assoc.), Gender (N. S.)

to be continued

Authors/Location	Method	Main Measurements	Main Findings about Guilt
Bohannon(1991b)/US	Xs./272 parents	Guilt:Sub - scale in GEI	R. F. :Attending church(- ve assoc.)
Miles and Demi (1991—1992)/US	Xs./132 parents	Guilt:Coding answers to the open questions about guilt	Prev. :In total: 83%, suicide bereaved parents: 92%, accident bereaved parents: 78%, Chronic disease bereaved parents: 71%
Hazzard, Weston, and Gutterres(1992)/US	Xs./45 parents	Guilt: One item on self - blame for the death & sub - scale in GEI; H. O. :Sub - scales in GEI	R. F. :Sudden death > Chronic illness; lost a boy > lost a girl; Self - blame & "Somatization": + ve assoc.; Self - blame & "Atypical reactions", "Despair", "Anger - hostility", "Loss of control", "Death anxiety", "Depersonalization": N. S.
Davis, Lehman, Wortman, and Silver(1995)/US	Study 1: Xs./40 parents & 50 spouses	Guilt: One item on perceived preventability; H. O. :SCL - 90 - R, CES - D; SADS	Prev. 58%; Guilt & psychological distress: + ve assoc.
	Study 2: Long. (T1: 3W, T2: 18M)/ 124 parents who lost infants in SIDS	Guilt: Two items assessing guilt emotion in ABS; H. O. :SLC - 90 - R	Guilt(T1) & psychological distress(both T1 & T2): + ve assoc.

to be continued

Authors/Location	Method	Main Measurements	Main Findings about Guilt
Weinberg(1994)/US	Xs./200 people	Guilt: One item on self-blame from RWCC; Grief: One item about recovery from the death	Prev. 46%; Self-blame & recovery from grief: – ve assoc.
Weinberg(1995)/US	Xs./200 people	Guilt: One item on self-blame from RWCC & One item about behavioral self-blame	Prev. 20% ~ 30%; R. F.: Unnatural death > Natural death; Self – blame & recovery from grief: – ve assoc.
Constantino and Bricker (1996)/US	Intervention efficacy investigation /32 spouse bereaved from suicide were randomly assigned to BGP or SGP	Guilt: Sub – scale in GEI; H. O.: BDI, BSI, SAS	Only BGP is effective to decrease guilt & anger, while both BGP & SGP reduce the psychological distress & other aspects in grief
Lang, Gottlieb, and Amsel (1996)/Canada	Long. (T1: 1 – 24M, T2: 2 – 4M)/57 (T1)/31 (T2) bereaved couples who lost their children & 57(T1)/36(T2) comparison couples	Guilt: Sub – scale in BEQ	R. F.: Time (– ve assoc.); Gender (women > men); Received support (– ve assoc. – for women)
Schwab(1996)/US	Xs./35 parents	Guilt: Sub-scale in GEI	R. F.: Gender(women > men)
Hyrkäs, Kaunonen, and Paunonen(1997)/Finland	Xs./318 spouses	Guilt: One item on survivor guilt in HGRC	R. F.: Time (N. S.); Gender (Women > Men, but only when bereavement time > 2Y)

to be continued

Authors/Location	Method	Main Measurements	Main Findings about Guilt
Bailley, Kral, and Dunham (1999)/Canada	Xs. /350 college students	Guilt: sub – scales in GEQ("guilt" & "responsibility")	R. F.: Suicide > other types on responsibility for death, but N. S. group difference on guilt; natural death > unnatural death on guilt
Field, Bonanno, Williams and Horowitz(2000)/US	Long. (T1:3 – 7M, T2:14M)/56 spouses	Guilt: Coding the monologue content & One item on self-reported guilt feelings; H. O.: TRIG, SCL – 90 – R	Coded guilt & grief: + ve assoc. (T1 & T2); Self-reported guilt & depression: + ve assoc. (T1), N. S(T2).
Gamino, Sewell and Easterling(2000)/US	Xs. /85 people	Guilt: Sub – scale in GEI; H. O.: HGRC	R. F.: Traumatic death(+ ve assoc.), younger age of the deceased(+ ve assoc.); Guilt & misery grief reactions: + ve assoc., personal growth: – ve assoc.
Field and Bonanno (2001)/US	Long. (T1:6M, T2:14M, T3:25M, T4:60M)/56 spouses	Guilt: Coding the monologue content & One item on self-reported guilt feelings; H. O.: TRIG, SCL – 90 – R; Continuing attachment: Self – constructed items	Concurrent association between guilt & grief: T1/T2/T3/T4(+ ve assoc.); T1 guilt predict grief at T2/T3/T4(+ ve assoc.); Guilt predict psychological distress at T4(+ ve assoc.); Maladaptive continuing attachment mediate the association between guilt & grief.

to be continued

Authors/Location	Method	Main Measurements	Main Findings about Guilt
Harwood, Hawton, Hope and Jacoby(2002)/UK	Xs. case controlled study /36 suicide bereaved people & 36 natural death bereaved people	Guilt: Sub – scale in GEQ	R. F.: Difference between suicide & nature death: N. S.
Boelen, van den Bout, and van den Hout (2003)/ The Netherlands	Xs. /329 adults	Guilt: Sub-scale(self-blame) in the GCQ; H. O.: ITG, SCL – 90	Guilt & grief reactions: + ve assoc.; Guilt & psychological distress: + ve assoc.
Boelen and Lensvelt – Mulders(2005)/ The Netherlands	Xs. /531 adults	Guilt: Sub-scale(self-blame) in the GCQ; H. O.: ITG, SCL – 90	Guilt & grief reactions: + ve assoc.; Guilt & psychological distress: + ve assoc.
Mizota, Ozawa, Yamazaki and Inoue(2006)/Japan	Xs. /Interview 46 & survey 307 family members of HIV – Infected Hemophiliac	Guilt: Self constructed five items; H. O.: self constructed items on grief reactions, IES – R, GHQ – 12	Prev.: 86% – 92%; R. F.: Guilt & grief reactions: + ve assoc.; Guilt & psychological distress: + ve assoc.
Surkan et al. (2006)/Sweden	Xs. /449 parents	Guilt: One item on guilt feeling	Prev. feeling of guilt monthly: 5%, weekly: 8%, daily: 11%; R. F.: Perception of receiving inadequate care(+ ve assoc.)
De Groot et al. (2007)/Netherlands	Intervention (CBT) efficacy in suicide bereavement /Intervention group (N = 68) & Control group(N = 54)	Guilt: self – constructed items	CBT intervention is effective to reduce guilt feeling

to be continued

Authors/Location	Method	Main Measurements	Main Findings about Guilt
Kowalski and Bondmass (2008)/US	Xs./173 widows	Guilt: Sub - scale in RGEI; H. O.: Self reported physical symptoms	R. F.: Time(N. S.); Guilt & physical distress: + ve assoc.
McMenamy, Jordan, and Mitchell(2008)/US	Xs./67 people	Guilt: One item on difficulties in dealing with guilt	Prev.: 73%
Torges, Stewart, and Nolen - Hoeksema(2008)/US	Long. (T1: 1M, T2: 6M, T3: 18M)/455 people	Guilt: Coding answers to an open question about guilt; H. O.: IDD	Unresolved regret(guilt)❶ & depressive symptoms: + ve assoc.; Interaction between regret (guilt) & time predict the decrease of depressive symptoms
Ylitalo, Valdimarsdottir, Onelov, Dickman, and Steineck(2008)/Sweden	Xs./379 widows	Guilt: One item on the guilt feeling	R. F.: Perception of receiving inadequate health care(+ ve assoc.)
Garnefski and Kraaij (2009)/The Netherlands	Xs./68 people	Guilt: Self - blame in CERG; H. O.: SCL - 90	Guilt & psychological distress: + ve assoc.
Akiyama, Numata, and Mikami(2010)/Japan	Xs./147 family members	Guilt: Self constructed three items; H. O.: 15 - GDS, self constructed items on grief reactions	R. F.: Time (N. S.), Peaceful death (- ve assoc.), Providing direct care to the deceased (- ve assoc.), fulfilled home care system(- ve assoc.); Guilt & grief reactions: + ve assoc.; Guilt & psychological distress: N. S.

❶ The content of "regret" in the study fits our definition of guilt in bereavement.

to be continued

Authors/Location	Method	Main Measurements	Main Findings about Guilt
Jonasson et al. (2011)/Sweden	Xs./907 widowers	Guilt: One item on the guilt feeling	R. F.: End of life discussion with the deceased(– ve assoc.)
Golden and Dalgleish (2012)/Croatia	Comparing Complicated Grief group (N = 20) with control group (N = 27)	Guilt: subscale (self – blame) in PTGI	Group difference on self – blame: N. S.

Note. BGP: Bereavement Group Postvention; H. O.: Health outcomes; Long.: Longitudinal design; M: Months; – ve assoc.: Significant Negative Association; N. S.: Non – significant; + ve assoc.: Significant Positive Association; Prev.: Prevalence; R. F.: Risk Factors; SGP: Social Group Postvention; SIDS: Sudden Infant Death Syndrome; W: Weeks; Xs.: Cross – sectional design; Y: Years.

Abbreviations for scales: ABS: Affective Balance Scale; BDI: Beck Depression Inventory; BSI: Brief Symptom Inventory; CBT: Cognitive Behavioral Therapy; CERG: Cognitive Emotion Regulation Questionnaire; CES – D: Epidemiological Studies Depression Scale; ESS: Experience of Stress Scale; GCQ: Grief Cognition Questionnaire; GEI: Grief Experience Inventory; GEQ: Grief Experience Questionnaire; GHQ – 12: the 12 – item version of General Health Questionnaire; HGRC: Hogan Grief Reaction Checklist; IDD: Inventory to Diagnose Depression; IES: Impact of Event Scale; ITG: Inventory of Traumatic Grief; PTGI: Posttraumatic Cognitions Inventory; RGEI: Revised Grief Experience Inventory; RWCC: Revised Ways of Coping Checklist; SADS: Scheduler for Affective Disorders; SAS: Social Adjustment Scale; SCL – 90/SCL – 90 – R: Symptom Checklist; STAI: State – Trait Anxiety Inventory; TRIG: Texas Revised Inventory of Grief; 15 – GDS: The 15 – item Geriatric Depression Scale.

Risk Factors

Who among the bereaved is more likely to suffer from guilt feelings than others, to an intensity that may be problematic, and what circumstances are associated with greater prevalence and intensities? Research in the guilt area has largely been restricted to the identification of correlates (rather than causal factors), that is, variables whose presence is associated with an increase in the likelihood of (problematic) guilt feelings, sometimes known as "risk factors" (W. Stroebe & Schut, 2001). The correlates investigated in relationship to guilt in bereavement can be grouped into (1) situational factors, such as the time since loss, cause of the death, characteristics of deceased, pre – death relationship and palliative care received; (2) personal factors, such as gender and religion; and (3) interpersonal factors, such as social support. It will become evident that some factors, such as gender, time, and cause of death have been researched more than others, but that, in general, there is a dearth of sound empirical evidence on correlates of guilt.

When it comes to the time course of guilt, longitudinal studies seem to provide stronger evidence compared with cross – sectional studies. Among the evidence we reviewed, guilt is found to diminish with time in longitudinal studies among bereaved parents (Bohannon, 1991a; Lang, Gottlieb & Amsel, 1996), yet such a pattern has not been identified in cross – sectional studies among bereaved spouses (Kowalski & Bondmass, 2008; Zisook & Shuchter, 1985) or among diverse subgroups of bereaved people (Akiyama, Numata, & Mikami, 2010; Mizota et al., 2006). Whether the time course of guilt differs across subgroups or countries is open to further exploration.

There seems to be a general assumption that unnatural death is a risk factor for intense guilt (Rando, 1993; W. Stroebe & Schut, 2001). However, it is hard to draw such a conclusion from the empirical studies, as results are highly discrepant. There are some studies reporting higher guilt in natural compared with unnatural death (Bailley, Kral, & Dunham, 1999). Other studies conclude that there are higher rates following unnatural than natural death (Gamino, Sewell, & Easterling, 2000; Hazzard, Weston, & Gutterres, 1992; Weinberg, 1994). Yet others report no differences across different types of death (Harwood, Hawton, Hope, & Jacoby, 2002). Suicide survivors have been expected to be more vulnerable to guilt than following other types of bereavement (Jordan, 2001). However, a review of well – controlled studies comparing this type of death with others found little evidence to support this assumption (Sveen & Waly, 2008).

Characteristics of the deceased, such as gender and age, have also been reported to influence guilt. However, it is likely that those results are influenced by other confounding variables. For example, one study found parents who lost a son experienced higher guilt than those who lost a daughter, but at the same time, more boys died suddenly than girls in this sample (Hazzard et al., 1992). Another study reported younger age of the deceased as a risk factor for guilt felt by family members (Gamino et al., 2000). This could be due to the confounding effect of cause of death too, as it is more likely for young people to die from unnatural causes than it is for the elderly.

More guilt also occurred when there had been an ambivalent pre – death relationship with the deceased, which is characterized by argument, dissatisfaction and frustration in the relationship (Shanfield & Swain, 1984). Several studies suggested the importance of a "good ending" to alleviate the risk of guilt. In cases where the deceased passed away peacefully, end of life discussion about the death or after death issues were carried on with the deceased, or adequate medical care was provided, then lower rates or less intensive guilt has been found among family members after the death (Akiyama et al., 2010; Jonasson et al., 2011; Surkan et al., 2006; Ylitalo et al., 2008).

Several studies suggest that women experience—or at least, express—more guilt than men. Three studies have compared guilt among wives and husbands who had lost a child (Lang et al., 1996; Mizota et al., 2006; Schwab, 1996). Despite the limited number of studies, the findings that mothers are at higher risk of guilt than fathers seems to be persuasive, particularly because these studies eliminate other confounding variables (such as duration of bereavement, relationship to the deceased) in the comparison. Nevertheless, it remains unclear whether the reported difference reflects different levels of guilt actually experienced by women and men, or just the tendency that women are more expressive about their negative emotions (Simon & Nath, 2004).

Religion is assumed to be helpful in adjustment to bereavement (W. Stroebe & Schut, 2001; Wortman & Park, 2008). Only one study we reviewed explored the influence of religion on guilt, and found an inverse relationship between attending church and guilt (Bohannon, 1991b).

Social support may be another protective factor. Lang et al. (1966) found that intimacy of the marital relationship (sharing with and receiving support from the partner) was associated with lower guilt among bereaved mothers.

Based on the limited number of studies conducted to explore correlates of guilt, it seems that women report more guilt than men, and tentatively, one can probably conclude that guilt tends to decrease with time, even though it is hard to determine the precise course of this decline. The unclear patterns regarding guilt following different causes of death may suggest that guilt is suffered across all subgroups of survivors, although it seems likely that it would be higher for some vulnerable groups than others. On the other hand, a good ending of life seems helpful in lowering the risk of guilt in bereavement. Commonly – accepted helpful factors in adjustment to bereavement, such as religion and social support, have received little exploration in relationship to guilt, probably due to the fact that grief is usually examined as the outcome variable and few studies have selected out guilt for separate investigation.

Guilt and Health Outcomes

There are studies exploring the association between guilt and health outcomes, with the latter ranging from depression, psychiatric symptoms, physical problems, normal grief, and complicated grief to personal growth. Seven of them are longitudinal studies, and most of them have reasonable sample sizes. Most studies indicate the detrimental impact—or, at least, association—of guilt on or with health outcomes. For example, guilt is found to be concurrently associated with depression. Moreover, guilt early on in bereavement predicts depression at 13 months, 14 months, 16 months, 18 months, even as long as 60 months after the death (Davis et al., 1995; Field & Bonanno, 2001; Field, Bonanno, Williams, & Horowitz, 2000; Torges, Stewart, & Nolen - Hoeksema, 2008). Besides depression, associations between guilt and psychiatric problems as well as traumatic reactions have been documented (Mizota et al., 2006). Guilt is found to be associated with impaired physical health positive correlations have been reported between guilt and "somatization" among bereaved people (Hazzard et al., 1992; Kowalski & Bondmass, 2008). The association between guilt and greater maladjustment is reflected not only from the findings on psychological distress but also from the positive correlation between guilt and grief symptoms, either normal or complicated manifestations. For example, guilt is found to be concurrently associated with grief reactions (Akiyama et al., 2010; Field et al., 2000; Gamino et al., 2000; Mizota et al., 2006; Weinberg, 1994), and guilt has also been found to predict grief symptoms at 14, 18, 24, and 60 months (Field & Bonanno, 2001; Field et al., 2000; Torges et al., 2008). The positive correlation between guilt and complicated grief is also reported in some studies (Boelen & Lensvelt - Mulders,

2005; Boelen, van den Bout, & van den Hout, 2003). Lastly, the inverse relationship between guilt and positive outcomes after bereavement further confirms the association between guilt and maladaptive health outcomes (Gamino et al., 2000). The harmful influence of guilt on health outcomes seems apparently based on the result of most studies, despite a few exceptions, as shown in Table 2.1.

Some preliminary suggestions on the mechanism underlining the association are available from these studies. A variety of different explanations, which are not necessarily mutually exclusive, have been put forward: (1) Self-blame adds extra distress which depletes coping resources to deal with grief (Horowitz et al., 1984); (2) Guilt or self-blame impair one's sense of worthiness, which in turn contributes to depression (Weinberg, 1994, 1995); (3) Guilt in bereavement is correlated with maladaptive continuing attachment, which is claimed to be responsible for hindering adjustment (Field & Bonanno, 2001; Field et al., 2000); (4) Guilt may be correlated with maladaptive strategies (such as avoidance coping), thereby impeding adjustment (Boelen & Lensvelt-Mulders, 2005). However, these propositions have been insufficiently tested or supported by research data.

In conclusion, the positive relationship between guilt and maladaptive health outcomes has been reported in most studies. Depression as the independent outcome has been discussed more often than other type of health outcomes, and only several studies include complicated grief in the investigation.

Critical analysis and research gaps

Even though there have been some explorations and findings about guilt in bereavement as summarized above, the limitations of those studies are also evident. Two main obstacles that weake the previous findings and prevent further exploration on guilt in bereavement are: (1) the lack of clear conceptualization; and (2) the lack of a reliable scale of guilt in bereavement.

Lack of clear conceptualization of guilt in bereavement

Among the relevant studies, only a few have defined guilt or related concepts (e.g., self-blame) before their investigation (eg. Field et al., 2000). Most did not provide a definition at all, but if one examines the specific measurements adopted to assess guilt, it becomes evident that guilt has been variously considered, for example, as a coping strategy (Garnefski & Kraaij, 2009; Weinberg, 1994), a (group) of cognitions (Boelen et al., 2003; Davis et al., 1995; Downey, Silver, & Wortman, 1990), or an emotion which is part of grief (Harwood et al., 2002; Kowalski & Bondmass, 2008; Lang et al., 1996). It is clearly difficult to build on the knowledge of guilt in bereavement without a clear and consistent conceptualization.

Lack of a valid measure for bereavement guilt

As mentioned above, the cognitive evaluation can generate various contents. There are evidences in other domains that support the concept of multiple sources of guilt. For example, the Care Giver Guilt Questionnaire comprises guilt about doing wrong toward care recipients, guilt about not rising to the occasion as caregivers, guilt about self-care, guilt about neglecting other relatives, and guilt about having negative feelings towards other people (Losada, Márquez-González, Peñacoba, & Romero-Moreno, 2010). The Trauma Related Guilt Inventory contains one sub-scale about global guilt, one sub-scale of guilt distress, and 3 other sub-scales regarding different guilt cognitions (Kubany & Abueg, 1996). In the field of bereavement, several qualitative studies have suggested that bereaved people may feel guilty for various reasons, such as not preventing the death, not doing enough for the deceased, enjoying life again after the loved one's death and so on (Kerr, 1994; Smith et al., 2011). Miles and Demi (1983, 1991, 1992) proposed a model of sources of guilt in bereavement from her study on bereaved parents. Her model includes death causation guilt, illness-related guilt, childrearing guilt, moral guilt, survival guilt, and grief guilt.

However, most previous studies in this field failed to adopt appropriate measures to reflect such different sources of guilt. The studies which constructed just one or several items to measure guilt (De Groot et al., 2007; Surkan et al., 2006), or adopted one or several items from an instrument which was not specifically designed to measure guilt (Weinberg, 1994) lack psychometric validity and only assess particular sources of guilt. Some studies collect qualitative data and code it to reflect guilt (Field & Bonanno, 2001; Field et al., 2000; Torges et al., 2008). This method opens the opportunity to explore all sources of guilt reported by participants. However, it does not delineate different components in guilt but simply sums up all sources to represent guilt in total. Additionally, this depends on the coder's understanding and definition of "guilt". Other studies use just one general item (e.g., "Did you feel guilt or self-reproach regarding the death?") to measure guilt (Jonasson et al., 2011; Ylitalo et al., 2008). This leaves the definition of guilt open to the participants, thereby not enabling one to eliminate subjective bias in interpretations, not to mention failing to identify separate, different sources of guilt. There are some sub-scales in bereavement scales measuring guilt, responsibility or self-blame (e.g., GEI, GCQ). Those sub-scales contain more than one source of guilt, but there is no evidence to show that they capture the comprehensive content of guilt. Again, such a strategy cannot be used to gain understanding of distinct aspects.

Why do we need a valid measure that can capture the multidimensional characteristics of guilt in bereavement? Apart from the fact that the reviewed studies have shown guilt to contain different components and sources, there are problems caused by poor measurement, and they limited understanding of this emotion. For example, the difference between high and low guilt prevalence in a study may be caused by the use of different measures which contain many aspects of guilt versus those which contain only one aspect. Furthermore, different sources of guilt may correlate with different risk factors. For example, one study found that the cause of death does not influence how much family members feel they did not do enough for the deceased, as measured by the guilt sub – scale in the GEQ, but it did show an effect on the family's perceived responsibility for the death (Bailley et al., 1999). Another study found that family members' self – blame contributed to depression beyond their perceived responsibility for the death (Davis, Lehman, Silver, Wortman, & Ellard, 1996). Together, such evidence suggests that different types of guilt contribute to health outcomes beyond self – blame for the death alone. The respective association between different sources of guilt and health outcomes can only be assessed with a tool that delineates guilt into sub – components.

In conclusion, the data strongly suggest multiple sources of guilt despite the fact that our understanding about the scope of these sources is still sparse at this time. A valid instrument which embraces the complexity and multi dimensional nature of guilt is important for future empirical investigation.

Existing conceptualizations and measures of guilt in psychology

Inconsistency in conceptualization

The difficulty of inconsistent conceptualization of guilt is a problem that exists in general psychology. A recent major review identified 23 theory – based definitions and 25 measures of guilt(Tilghman - Osborne, Cole, & Felton, 2010). They coded definitions based on 18 characteristics, such as developmental differences, moral transgression/social transgression, self - focus/no self, behavior, public/non public, trait - like, situation specific and so on. The results found that there was general lack of conceptual convergence, as different conceptualizations shared only 60.7% of the same codes on average. Several differences in the conceptualization are worthy to be highlighted for consideration in defining guilt in bereavement.

Trait or state. Some theories consider guilt as a personality characteristic (Kochanska, Gross, Lin, & Nichols, 2002), some define guilt as a transient emotional experience(Tangney, 1999), and others think guilt can be both a trait - like and state - like construct(Harder & Zalma, 1990).

Moral or social transgression. Some definitions of guilt involve only moral transgressions(Kugler & Jones, 1992), others only relate to social transgression (Alexander, Brewin, Vearnals, Wolff, & Leff, 1999), while most definitions include both(Harder & Zalma, 1990).

Evaluation on behavior or self. In some definitions of guilt, a negative evaluation of self and identity is evident(R. C. Johnson et al., 1987), but other definitions specifically state that the object of the evaluation is a person's behaviors and activities(Tangney & Dearing, 2002).

Subtypes. In some theories, guilt is not delineated into separate subtypes (Kugler & Jones, 1992), whereas other researchers have proposed different kinds of guilt, referring to certain events; such as survivor guilt, separation guilt, omnipotent responsibility guilt and self - hate guilt (O'Connor, Berry, Weiss, Bush, & Sampson, 1997).

Problems in measures

After analyzing 25 measurements of guilt, the review(Tilghman - Osborne et al. ,2010) also indicated that there is lack of consistency between theories of guilt and correspondence measures. Some guilt scales measure characteristics not stated in their definitions, while others omit important features in their definitions. For example, according to Tangney and her colleagues, guilt is defined as a bad feeling arising from a negative evaluation of a specific behavior, which behavior is found to be immoral, lacking or otherwise defective (Tangney, Burggraf, & Wanger, 1995). However, the corresponding measurement assesses behaviors or behavioral intentions in specific events or situations (Tangney, 1990).

In addition, some measurements contain constructs which are not elements of guilt, such as depression, sadness or worry (O'Connor et al. ,1997). This impairs the scale validity and adds a confounding effect when assessing the relationship between guilt and psychological outcomes.

The trouble: discrepant findings of guilt and mental health outcomes

There has been a lot of research evidence to support the detrimental role of guilt in mental health. Some psychologists believe that depression derives from exaggerated guilt (Murphy, 1978), and the Diagnostic and Statistical Manual of Mental Disorders (DSM – IV) includes inappropriate and excessive guilt as one of the symptoms associated with major depressive disorder (APA. 1994). Higher guilt are found among depression patients than among a normal comparison group of people (Gupta, Kulhara, & Verma, 1988; Jarrett & Weissenburger, 1990). The research conducted in non – clinical samples has also revealed supporting evidence of the positive correlation between guilt and depression (Bybee & Zigler, 1996; Harder, Cutler, & Rockart, 1992; Harder & Zalma, 1990; O'Connor et al. , 1997). The symptomatic role of guilt in depression is found in children as well. A group of preschool children were interviewed to assess their depression symptoms, and their guilt experiences were collected by means of their parents' reports as well as through their narratives in a story telling task. The findings illustrate a positive correlation between children's guilt experiences and the severity of depression, and such association existe in children as young as 3 years old (Luby et al., 2009). Similarly, guilt is also reported to be associated with anxiety disorder and play a crucial role in developing and maintaining obsessive – compulsive disorders (Harder et al., 1992; Harder & Zalma, 1990; Lowe, 1964; Shapiro & Stewart, 2011).

However, discrepant findings about the relationship between guilt and psychopathology are found in the literature as well. Some studies did found no significant correlation between guilt and anxiety (Nixon & Steffeck, 1976), and others reported no significant correlation between guilt and depression or anxiety, either among the normal population or among patients with depressive disorder (Fergus, Valentiner, McGrath, & Jencius, 2010; Gojowy et al., 2011). Guilt has also been found to positively correlated with positive well – being (Orth, Robins, & Soto, 2010).

The inconsistent conceptualization and problematic measures probably account for the discrepant findings between guilt and psychopathology (Tilghman – Osborne et al., 2010). This signifies the necessity and importance of having a proper conceptualization along with a reliable measurement when studying guilt.

Developing a domain specific understanding of guilt

Without a consistent definition and an accordingly developed valid measure, it is difficult to accumulate understandings on guilt, as findings in different studies are hardly comparable. Given the situation of defining and measuring guilt in general psychology, it seems inappropriate, or even impossible to adopt an existing definition and scale to use in a study on guilt in bereavement. In fact, conducting domain specific research on guilt is advocated by researchers (Tangney, Stuewig,

& Mashek, 2007), since "guilt requires a situational context for its manifestation" (Tilghman - Osborne et al., 2010, p. 544), and it is important to explore "for whom and under what conditions does guilt become maladaptive" (Tangney, 1999, p. 553).

Guilt has been investigated in some domains where guilt is commonly reported. For example, guilt related to trauma has attracted more and more research attention. Clinicians and researchers have noticed this common feeling among trauma survivors, and empirical efforts have been made to conceptualize trauma-related guilt and to develop a valid measurement (Kubany, 2003; Kubany & Abueg, 1996; Kubany, Abueg, Kilauano, Manke, & Kaplan, 1997; Opp & Samson, 1989; Pereda, Arch, Peró, Guàrdia, & Forns, 2011). Trauma researchers have been exploring the relationship between guilt and post trauma stress, and reveals a positive correlation between guilt and PTSD in survivors from various kinds of trauma; such as war, holocaust, terrorist attack and intimate partner violence (Beck et al., 2011; Browne, Evangeli, & Greenberg, 2012; Henning & Frueh, 1997; Mallimson, 2006; Owens, Steger, Whitesell, & Herrera, 2009). Based on the accumulated evidences, some researchers even propose to include guilt as one feature in PTSD (Beck et al., 2011), and there are studies have started to explore the underling mechanisms of such connections (Held, Owens, Schumm, Chard, & Hansel, 2011; Street, Gibson, & Holohan, 2005).

In conclusion, it is necessary to have a domain specific definition and measurement before accumulating more understanding of this phenomenon under this specific context. Therefore, based on the analysis of previous definitions of guilt in general, and previous empirical studies in bereavement, a conceptualization of bereavement guilt will be proposed before carrying on future investigations.

The proposed conceptualization of bereavement guilt

As mentioned earlier, psychologists have provided various definitions of guilt. However, by comparing and analyzing different theories of guilt, Tilghman - Osborne and Cole (Tilghman - Osborne et al., 2010) found several common features in various definitions (e. g. transgression), based on their proposal, guilt has both affective and cognitive components, involving "moral transgressions (real or imagined) in which people believe that their action (or inaction) contributed to negative outcomes…A sense of responsibility and painful feelings of remorse are part of the guilt" (p. 544). This suggestion is similar to the definition given by some trauma and bereavement researchers. For example, in the trauma field, Kubany (2003, p. 53) defined guilt as "an unpleasant feeling accompanying beliefs that one should have thought, felt or acted differently". Bereavement researchers have defined guilt in slightly different ways, for example, as the feeling that "results when one perceives that one has done something wrong, failed to live up to personal expectations, or violated a societal standard." (Miles & Demi, 1991—1992, p. 205), or as "… the feeling of culpability deriving from perceived offenses or a sense of inadequacy." (Rando, 1993, p. 478)

Two core elements in guilt that seem useful for our current purposes can be identified in the above definitions: the cognitive evaluation on one's own behavior and the feeling accompanying it. Such view of guilt is consistent with the nature of guilt as one type of self - conscious emotion, differentiating it from basic emotions, such as sadness or fear. While all emotions may be cognitive to some degree, the cognitive component is, however, an indispensable component in self – conscious emotion, which is developed out of self reflection and self evaluation (Tangney, 1999; Tracy & Robins, 2007). The conceptualization of guilt as containing both cognitive and affective components is agreed upon by different guilt researchers and theoreticians (Berrios, Bulbena, Bakshi, & Dening, 1992; Kubany, 2003; Kubany & Abueg, 1996).

Guilt in bereavement has been documented as one emotion in grief, and is incorporated as one aspect of grief in classic grief measures. Given this historical treatment and trying to incorporate the core elements of previous definitions, bereavement guilt refers in this research to:

A remorseful emotional reaction in grieving, with the recognition of having failed to live up to one's own inner standards and expectations in relationship to the deceased and/or the death.

Several points need to be stressed about this definition. First, bereavement guilt is conceptualized as an emotional experience in grief rather than a personality. Second, this construct contains both cognitive and affective components in accordance with its nature as a self – conscious emotion and in agreement with previous researchers. Third, this definition does not emphasize or exclude social or moral transgression, but instead stresses the role of inner standards. That is, the evaluation is subjective. Inner standards are a crucial basis for evaluation, even though they are influenced by social and moral standards. Fourth, it is understood that there can be multiple sources of guilt generated from cognitive evaluation. Therefore, in this conceptualization, it is supposed that bereavement guilt is a multi dimensional construct; or in other words, there may be sub – types of guilt within the bereavement construct. What those sub types are will be one of the research questions in this project.

Chinese Bereaved Adult Children

Bereaved adult children

The literature on bereaved people is more inclined to focus on bereaved young children, bereaved parents or bereaved spouses, even though bereaved adult children make up the largest proportion of bereaved people, considering the mortality rate among different age groups. Most people experience losing a parent(s) as an adult, rather than as a child. With the improvement of medical and health conditions, people are now enjoying longer life spans than before, more people live to an older age than was common decades ago. This in turn increases the number of people who have lost their parents at middle or even old age. According to a statistical report, only 10 percent of children under age 25 have lost one of their parents, while 50 percent of children have lost both parents by age 54 (Winsborough, Larry, & William, 1991).

Even though losing a parents is a typical experience for most adults, it is not necessarily an easy experience. Instead, it is one of the most distressful events that an adult might encounter, because of the significant relationship between children and their parents. Most of the time, the mother or the father is the primary attachment figure for the child (Bowlby, 1980). They provide a sense of warmth and security for the child, and the relationship is so significant that when children grow up, this attachment motivates them to protect their parents, in order to maintain this sense of support and security as being alive and still available to them (Cicerelli, 1993). Various researches have proven that the parents child relationship is significantly associated with adult children's well - being (Umberson,

1992). When such an important figure passes away, it is predictable that the bereaved adult children will suffer great distress. In fact, within the limited number of studies focusing on bereaved adult children, researchers have already suggested that losing a parents, even as an adult, is extremely distressful and can negatively influence a person's psychological as well as physical health (Marks, Jun, & Jieun, 2005). For example, in a survey taken from 220 bereaved adult children who lost their parents within the preceding 5 years, the results indicate that about 25% of the participants' social and emotional functions were severely impaired (Scharlach, 1991). Similar results were found in a longitudinal study, which compared bereaved adult children with a control group. Comparing with adults who had not lost a parents, the bereaved adult children experienced significantly increased psychological distress, alcohol consumption and decreased physical health status (Umberson & Chen, 1994). The qualitative analysis of bereaved adult children's responses also revealed unrecognized stress that is not expected by society, and how dramatically this event had changed their lives, as well as how little space has been provided to understand their experiences (Umberson, 2003).

Guilt in bereavement may be an especially meaningful phenomenon to this sub - group of bereaved people, because they are subject to a large number of duties and obligations to their parents. Those obligations are prominent in Chinese society, as expressed by filial piety. Filial piety is a traditional Chinese value that controls intergenerational relationships and can be extended to apply to interpersonal relationships outside the family. It origins from Confucianism, and writings about filial piety can be traced back to as early as the Pre - Chin Era (551 to 221 B. C.); for example, The Analects (*Lun Yu*), and the Book of Filial Piety (*Xiao Jing*). Modern psychologists have analyzed this classic reference about filial piety and found 15 key attributes; such as obedience to parents, supporting parents and protecting parents (Yang, Yeh, & Huang, 1989). Following the analysis of these elements, some researchers have proposed two main factors; namely, reciprocal filial piety and authoritarian filial piety. The former comprises the values of respecting and loving parents, and supporting and memorializing parents. The latter comprises suppressing oneself to comply with and defer to one's parents' wishes, as well as glorifying one's parents (Yeh & Bedford, 2003). Similar aspects, such as obedience, respecting and supporting one's parents, and continuing the family line, are also suggested by other researchers (Ho, 1996).

Taken together, bereaved adult children are a large group of people who have experienced the considerable influence of losing their parents even as adults. There is a dearth of understanding of the grief experience of this group of people within previous literature.

Cultural differences on guilt

The importance of cultural sensitivity has been acknowledged by bereavement researchers (Klass & Chow, 2011; Rosenblatt, 2008) and emotional psychologists (Markus & Kitayama, 1991), yet most studies about guilt in bereavement have been conducted in western countries. The importance of culture context to this topic can be illustrated from three aspects.

First, guilt is an emotion based on self evaluation against inner standards which are influenced by moral codes and cultural beliefs and conventions. However, the moral system in eastern countries is different from that in western countries. In western countries, liberalism is the moral – political philosophy that serves as the foundation for individualism. In contrast, Confucianism in eastern countries serves as the foundation for collectivism (U. Kim, 1994). In individualistic cultures, one's identity is viewed as being independent, and the societal moral code is based on protecting individuals' rights and freedom. Therefore, regulations, rules and laws are implemented in western countries to stress "negative duties", such as

"Do not kill". In western cultures, failing one's negative duties will be condemned, whereas positive duties are viewed as virtues and as being up to personal choice. In collective cultures, the self is viewed dependently, and a person's identity is defined within the relationship. Priority is given to the group rather than to the individual. A virtue – based moral system emphasizes duties and obligations to promote collective warfare and societal harmony (Bedford & Hwang, 2003; Markus & Kitayama, 1991). Therefore, positive duties (e. g. "respect parents") are mandatory in eastern cultures. If one fails the duties and obligations ascribed to one's role, that person will feel the risk of losing his/her group status, which is important in defining the identity of self (Bedford & Hwang, 2003; Hwang, 2001). Filial piety is just one such kind of positive duty in Chinese culture.

Second, negative self – conscious emotions, such as guilt and shame may be more significant in Chinese society than in western societies. Guilt is considered by Chinese people to be one of the basic emotions under the category of "shame", while this is not the case among participants from western countries (Shaver, Wu, & Schwartz, 1992). The hypercognition of shame and guilt also supports the idea that these emotions are more elaborate in Chinese culture than in the west (Russell, 1991). For example, in Chinese, there are words to describe different types of shame and guilt for which there are no corresponding translations in English that can convey the subtle nuances in meaning (Bedford, 2004). In fact, there are over 100 words related to guilt and shame in Chinese, while far fewer words describing these concepts can be found in English (Jin Li, Wang, & Fischer, 2004; Shaver et al., 1992).

Third, the tendency to experience guilt, as well as the relationship between guilt and pathology, may differ across different cultures. Some studies found that participants in eastern countries are more prone to guilt compared with people from western countries (Anolli & Pascucci, 2005; Bear, Uribe – Zarain, Manning, &

Shiomi, 2009). Guilt is found to be less associated with negative consequences in collective culture compared to individualistic cultures (Wallbott & Scherer, 1995).

Taken together, culture differences may exist in both the elicitors and the prevalence of guilt in bereavement, as well as in its relationship to grief outcomes. Therefore it is necessary and important to observe the Chinese cultural context when investigating this phenomenon.

The present research

In summary, guilt is a common emotion in grief, and it is likely to associate with pathological adjustment in bereavement. Previous studies have made the effort to explore the prevalence and risk factors of guilt, as well as its relationship with grief outcomes. However, deficiencies in conceptualizing and assessing guilt cast doubt on the interpretation of those findings, and also prevent further in-depth understanding. Based on the analysis of key components in guilt and in the historical treatment of guilt in bereavement, a new definition of bereavement guilt is therefore now proposed in the present research. Guilt in bereavement is considered to contain both cognitive and affective components, and involves various contents within guilt cognition. The research objective is to consolidate a sound conceptualization of guilt and to build a valid measurement, in order to enhance researchers' and care professionals' understanding of guilt among Chinese bereaved adult children.

Specifically, it aims to

a) build up a conceptual model of this phenomenon,

b) develop a valid measurement,

c) and implicate the conceptualization and measurement to further explore the understanding of this phenomenon on the group level, including its prevalence, risk factors and association with complicated grief.

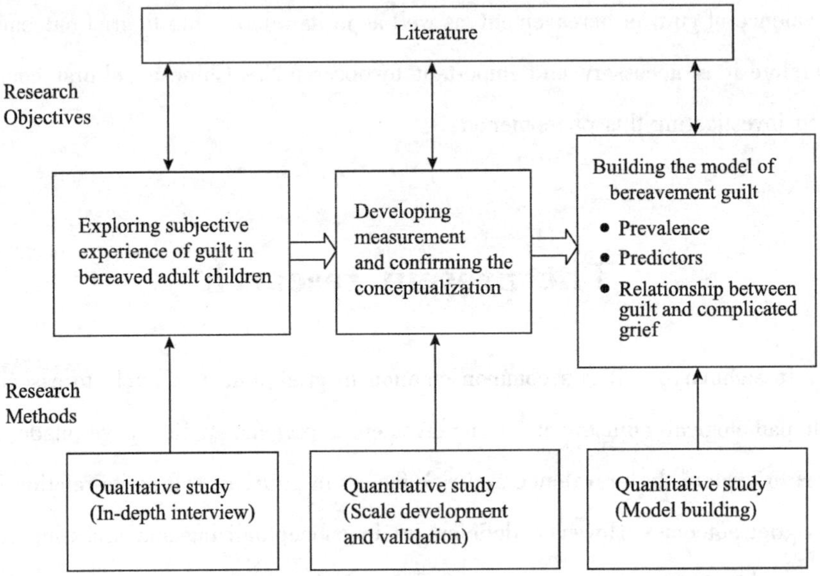

Figure 2.1 Research Framework

In order to reach those objectives, I will first adopt a bottom up inquiry method to let Chinese bereaved adult children speak for themselves about their experience of guilt. After consolidating the definition and conceptualization of guilt based on the information collected from those persons with relevant experience, a measurement will be developed and validated accordingly. This tool will be utilized to address the discrepant findings in previous literatures, such as the enormous range of guilt prevalence among bereaved people. It is also helpful to explore demographic and death related risk factors of guilt in bereavement in order to provide information regarding which group of people is most at risk. Finally, the relationship between guilt and complicated grief can be assessed. This not only extends the evidence of the role of guilt in psychopathology into the bereavement domain, but also provides information on the etiology of complicated grief and will shed light on clinical interventions.

The research framework of the present research is illustrated in Figure 2.1.

Chapter 3

Methodology

As introduced in the last chapter, there are multiple research objectives in this project. In order to meet the research objectives and provide valid answers to research questions, adopting appropriate research methods plays a crucial role. The research questions raised by this project cannot be answered by using only either quantitative methods or qualitative methods, so a mixed – method approach is appropriate. This chapter will introduce the rationale for using a mixed method design, the definition and advantages of this approach, the paradigms utilized to accommodate the approach, and the methodological framework for the present research program. It will provide an overview of the methodology issue underlining the whole research project, while specific data collection and analysis procedures in each study will be elaborated upon in the following chapters.

Meeting research objectives

The research objective determines research questions to be asked (Newman, Ridenour, Newman, & Jr. GeMarco, 2003), and "research methods should follow research questions in a way that offers the best chance to obtain useful answers" (R. B. Johnson & Onwuegbuzie, 2004, p. 18). Both the quantitative method and the qualitative method have their strengths and weaknesses in answering research questions (R. B. Johnson & Onwuegbuzie, 2004), and certain research questions are best asked by using specific methods (Sale, Lohfeld, & Brazil, 2002). Meeting the research purpose therefore, should be the dominant consideration when choosing an appropriate research method.

To better clarify the rationale behind a mixed method design, it is necessary to restate the research purpose and research questions here. The present research program aims to enhance researchers' and care professionals' understanding of guilt among Chinese bereaved adult children. Specifically, it aims to conceptualize this phenomenon, develop a valid measurement, and implicate the conceptualization and measurement to further explore the understanding of this phenomenon on group level, including its prevalence, demographic and death related predictors, and its association with complicated grief.

Even though there are various definitions of guilt in the field of social psychology, there is no developed theory or conceptualization of this emotion in the domain of bereavement, not to mention within the specific subgroup of the Chinese culture. Therefore, the conceptualization of bereavement guilt is best informed by the bereaved adult children's subjective experience of guilt. A qualitative research method with an inductive paradigm will allow the insiders' voices to be heard, and provide in-depth descriptions for conceptualization and for developing an assessment tool. The measurement will then be developed based on the qualitative findings and be validated through quantitative data in order to ensure its psychometric properties. Finally, a population level understanding and its relationships to other variables should be explored by means of a quantitative survey, so that we can discover how frequently this phenomenon is experienced by Chinese bereaved adult children in exact numbers, and so that its relationship to relevant correlates can be tested statistically.

It can be concluded from the above analysis that both qualitative and quantitative approaches need to be adopted to answer the different research questions and achieve all the research aims. Therefore, the mixed method design emerges as an optimal methodological option.

What is mixed method research

The emergence and development of mixed methodology has been suggested to be the third principal methodological movement in the social and behavioral sciences, following the dominance of quantitative method in the first half of the 20th century and the revolution of qualitative method in the past three decades (Teddlie & Tashakkori, 2003). Actually researchers have been adopting both qualitative and quantitative methods to answer their research questions for a long time, and the increasing popularity of mixed method design has prompted a specific journal (Journal of Mixed Method Research) to facilitate its discussion and development. However, it is difficult to give a unique answer to the question, "What is mixed method research?", because researchers have assigned various names to it, and diversities exist among different definitions (Teddlie & Tashakkori, 2003). Based on an recent updating of leading experts' views in the field, Johnson, Onwuegbuzi and Turner (2007) offered the following as a general definition:

"*Mixed method research is the type of research in which a researcher, or team of researchers, combines elements of qualitative and quantitative research approaches (e. g., the use of qualitative and quantitative viewpoints, data collection, analysis, inference techniques) for the broad purposes of breadth and depth of understanding and corroboration.*" (*R. B. Johnson et al.*, 2007, *p.* 123)

Moreover, the ideas of mixed method study and mixed method program have been separated further. The former refers to mixing in a single study, while the latter refers to a research program comprising a series of studies, each adopting different research methods (R. B. Johnson et al., 2007). This project belongs to the mixed method program category.

Advantages and functions of mixed method

Apart from the requirements imposed by specific research questions, a cluster of advantages and functions of mixed method research have been suggested by methodologist (for a review, see R. B. Johnson et al., 2007). Referring to the five most cited rationales by Greene, Caracelli, and Graham (Greene, Caracelli, & Graham, 1989), in the present research program, mixed method can serve the purpose of complementarity (enhancement of the findings in qualitative study from exploratory factor analysis and confirmatory factor analysis in scale development and model testing parts), development (using the results from qualitative study to help inform scale development), and expansion (expanding the breadth and range of understanding of bereavement guilt, in terms of the subjective experiences of adult children, the dimensions and structure of this concept and its association with other factors of interest, by using both qualitative and quantitative research methods). Three of the four recent rationales suggested by Collins, Onwuegbuzie, and Sutton (2006) can also be applied here. They are as follows: participant enrichment (optimizing the sample using specific techniques), instrument fidelity (assessing the appropriateness of existing instruments and creating new instruments), and significance enhancement (facilitating the thickness and richness of data). One of the most important utilizations of the mixed methodology in the present research program is for the development of a measurement, in which the significance of combining both qualitative and quantitative methods has been widely acknowledged by researchers (Bryman, 2006; Greene et al., 1989; Onwuegbuzie, Bustamante, & Nelson, 2010).

Paradigm debate and pragmatism

The strength of combining qualitative and quantitative methods has been widely accepted by researchers, and substantial work has been conducted on the technical level. However, there is still an on – going debate on the philosophical or paradigmatic level, which comprises a patterned set of assumptions on ontology (what is reality), epistemology (what is the knowledge of reality), and methodology (ways to know reality) (Guba, 1990).

The quantitative paradigm is based on positivism, which presupposes that the phenomenon is represented by indicators that can reveal the truth (Willig, 2001). This is based on the ontological position of singular truth and objective reality, and is usually deductive. The results of quantitative findings can test the relationship between variables, and uses randomization, blinding, and structured tools and statistics to generalize the findings to the population. Qualitative methods are embedded in interpretivism, hermeneutics and constructivism (Lincoln & Guba, 2000). This philosophy posits multiple realities and multiple truths, which are constructed in the minds of individuals. One can never approach reality without influencing it, or without it being shaped by the context or situations present. Purposive samples are used in qualitative studies to inquire after important information, but not to represent a larger group. This method tends to be induction driven.

In order to integrate both qualitative and quantitative research strategies, an alternative paradigm is needed to accommodate the different natures of the study, and pragmatism is the method most commonly advocated (Feilzer, 2010). Pragmatism is a standpoint from which to view the world with different layers and elements, and knowledge is viewed as being both constructed and based on the reality of the lived world(Dewey,1925). It rejects traditional dualism, reiterating that a philosophical stance should be judged based upon how well it works to solve problems. It argues that the aim of research is not to discover what is either accurate or relative about reality, rather, it is about what is useful, and for whom(Rorty,1999). Morgan(2007) has made a concise comparison between the pragmatic approach and other approaches, as shown in Table 3.1.

Table 3.1 A Pragmatic Alternative to Key Issues in Social Science Research Methodology

	Qualitative Approach	Quantitative Approach	Pragmatic Approach
Connection of theory and data	Induction	Deduction	Abduction
Relationship to research process	Subjectivity	Objectivity	Intersubjectivity
Inference from data	Context	Generality	Transferability

Note. Adapted from " Morgan, D. L. (2007). Paradigms lost and pragmatism regained: Methodological implications of combining qualitative and quantitative methods. *Journal of Mixed Methods Research*, 1(1), 48-76."

According to this view, the pragmatic approach is abductive, by which approach the researcher moves back and forth between induction and deduction. The concept of intersubjectivity maintains the possibility of the co-existence of both a single real world and a unique world to distinct individuals. Lastly, the transferability of the pragmatic approach argues that it is more important for investigators to decide to what extent the findings of a study can be transferred to other settings and what factors will influence this transferability, than to determine whether it is a context bonded or generalized finding.

The pragmatic stance provides a philosophic framework from which to integrate qualitative and quantitative elements. It acknowledges multiple layers of phenomenon, as well as the utility of the study. This fits well with the present research program, which aims to explore different aspects of bereavement guilt.

Methodological Framework

The present research program comprises several studies, each of which asks specific research questions to complete all of the research purposes together. Guided by Johnson and Onwuegbuzie (R. B. Johnson & Onwuegbuzie, 2004), two questions need to be considered when constructing a mixed - method design: (1) whether the research will be operated under one dominant theoretical drive or not, and (2) whether the mixing happens concurrently or sequentially. In the present research program, both inductive and deductive driven methods will be emphasized, and the studies will be conducted sequentially. It is a *Equal Status Sequential Mixed Method Research Program* (R. B. Johnson & Onwuegbuzie, 2004; R. B. Johnson et al., 2007).

The methodological framework of the research design is presented in Table 3.2 to illustrate the match between research purpose, research questions, research method and theoretical mode. The qualitative study will be conducted first to explore participants' subjective experiences, after which the results will then inform the conceptualization and be translated into items for scale development. Then, the conceptual model will be built up through exploratory factor analysis and the scale will be validated. Next the proposed structure of bereavement guilt will be tested and compared with other competing structure models in confirmatory factor analysis. Finally, the conceptualization of bereavement guilt and the constructed valid measure will be applied to explore the nature and consequences of bereavement guilt in Chinese bereaved adult children.

Table 3.2 Methodological Framework

	Research Purpose	Research Question	Research Method	Theoretical Drive
Study 1	Identify themes and narratives for conceptualizing bereavement guilt	What is the subjective experience of bereavement guilt in Chinese adult children?	Qualitative	Inductive
	Scale development and validation	What is the factor structure of bereavement guilt?	Quantitative	Inductive
Study 2		How good are the psychometric properties of the developed Bereavement Guilt Scale?		
	Test the structure of the concept	Is the factor structure of bereavement guilt supported by empirical data?	Quantitative	Deductive
	Build up the model showing the nature and consequences of bereavement guilt	How frequently is bereavement guilt experienced by Chinese adult children?		
Study 3		What demographic and death - related factors predict higher bereavement guilt in Chinese adult children?		
		What is the association between guilt and complicated grief?	Quantitative	Deductive

In summary, this chapter has discussed the methodological issues of the whole research program, which adopts a mixed method design, after considering the requirements related to the research purposes. Under the pragmatic paradigm, different studies will chose suitable research methods to collect and analyze data, and these procedures will be elaborated in chapter 4,5 and 6 respectively.

Chapter 4

The subjective experience of guilt: Insiders' voices

Introduction

Even though guilt has been reported as a common emotional experience in grief, there has not been a clear conceptualization of this phenomenon in the area of bereavement, and little is known about the experience of bereavement guilt. Therefore, adopting a bottom up paradigm using a qualitative approach through which to listen to the insiders' own voices relating to their experiences is an optimal starting point from which to begin the exploration. The qualitative exploration process can help the researcher to traverse descriptive inquiry, seeking to understand how people experience their guilt in bereavement. It is also an useful approach by which to understand how respondents make sense of their lives (Massey, Cameron, Ouellette, & Fine, 1998), thus it has been acknowledged as an important approach by which researchers can understand bereaved persons' grief experiences (Gillies & Neimeyer, 2006).

Early qualitative studies on this topic are from clinical case reports, which describe clients' symptoms believed to originate from guilt after losing someone they loved (Berman, 1978; Huang & Tang, 1996; Tooley, 1975), and suggested that treatments dealing with guilt are effective in reducing related somatic symptoms reported by the bereaved (Zimmerman, 2001). One interview study found that guilt was one of the main motivators for bereaved family members to sue the doctors who had treated the deceased (Chiu, 2010). A law suit is a way for bereaved adult children to release their feelings of guilt. More qualitative studies have gradually emerged to identify guilt related experience as one of the main themes in people's bereavement experience (Adolfsson & Larsson, 2004; Berman, 1978; Cacciatore, 2010; Hasui & Kitamura, 2004; Kerr, 1994; M. E. Smith, Nunley, Kerr, & Galligan, 2011; Tooley, 1975). Some bereavement researchers have made the effort to understand the cognitive content of guilt, and have proposed a model demonstrating sources of guilt among bereaved parents (Miles & Demi, 1983, 1991, 1992).

However, the findings from these studies have not provided enough valid information to conceptualize bereavement guilt due to following limitations.

First, most of these studies were focused on grief and bereavement experience, and guilt was only found to be one of the emerging themes, rather than being the main focus of the exploration. Guilt is identified as one aspect of grief reaction in these qualitative descriptions, and thus there is a lack of initiative and descriptions related to the components or content of this phenomenon.

Second, the theme of guilt identified in the above mentioned studies mainly concerns the cognitive component of guilt. In another word, only certain clues about the reason that people feel guilt can be inferred from these studies, while the affective component has never been elaborated upon or explored. Even though guilt is defined as one kind of emotion, people's subjective experiences related to the feeling component is missing in previous qualitative inquiries.

Third, most of these studies were conducted within western cultural environments. Only a few of them mentioned this phenomenon among Chinese bereaved persons (Chiu, 2010; Huang & Tang, 1996; Woo & Chan, 2010). Even in those Chinese studies, information about the subjective experience of guilt was not elicited, and neither were the cultural characteristics of the guilt experienced by Chinese people analyzed. As stated in the literature review, the cultural context is a significant factor that should not be ignored when investigating this topic.

Fourth, few previous qualitative investigations were conducted among bereaved adult children. However, exploring guilt among bereaved adult children is especially meaningful in the Chinese context, given the significance of the relationship and the influence of filial piety, which imposes a large number of role obligations upon children toward their parents(Ho, 1996).

Objectives

This study is one part of a bigger qualitative project which aims to understand the phenomenological experience of guilt among Chinese bereaved adult children[1]. In this study, I focus only on information about participants' subjective experiences, and these reveal themes that should be included when defining bereavement guilt. The narratives of the participants will be analyzed in order to:

(1) enrich the understanding of bereavement guilt and form a preliminary conceptualization, and

(2) build up an items pool for scale development in the next study.

Therefore, only findings that can help conceptualize bereavement guilt will be collected and analyzed in this chapter.

Method

Participants

Sampling

The principal sampling strategy in qualitative research is to find "information – rich cases", from which "one can learn a great deal about issues of central impor-

[1] The phenomenological experiences in the big project include subjective experience of guilt in bereaved adult children, how do they cope and live with it and how does guilt influence their adjustment in bereavement.

tance to the purpose of the research" (Patton, 1990, p. 169). Therefore, in the present study, a purposeful sampling method was adopted to recruit participants who can provide greater information about their phenomenological experience of guilt after their parents passed away. Maximum variation sampling, which is one of the various purposeful sampling method strategies, was adopted in the present study. It concerns participants' variations within some characteristics, while at the same time analyzing the significant themes that emerge from out of heterogeneous individuals. However, it does not aim to provide generalized conclusions, but to search for in – depth and diversified understandings within the specific variation. Several demographic aspects were outlined before recruiting, attempting to ensure at least one participant in each condition listed in Table 4.1.

Table 4.1 Characteristics Variation in Sampling

Characteristics	Condition
Gender	Male/Female
Age	18 – 30/31 – 50/Above 50
Lost parents	Only father/Only mother/Both parents
Bereavement time	Within 1 year/1 – 3 year/Above 3 years
Cause of death	Natural death[a]/Un – natural death[b]

Note. [a]: including illness related death, or natural organ failure; [b]: including other causes of death which is not included in the type of natural death.

In order to provide information related to the research interests, I looked for adult Chinese whose parents passed away when or after they were 18 years old. Meanwhile, the person must have experienced or is still experiencing guilt in his or her grief experience. Those who have been diagnosed with psychotic disorders or with cognitive impairments were excluded, in order to protect them from possible psychological distress they might incur from the interview, and to ensure the quality of communication.

Recruitment

The recruiting process began with sending out a recruiting advertisement via the researcher's personal social network, and asked them to spread this information to potential participants that they might know. Meanwhile, the recruiting advertisements were disseminated through talks, workshops and symposiums on related topics held by the Center on Behavioral Health (CBH) in the University of Hong Kong. Attendants in the above activities were mainly bereaved people from within the community, social workers, counselors and nurses who have worked or are working with bereaved persons. The recruiting information was expected to reach potential participants though their dissemination. Contact information was left in the advertisement, and when a potential participant contacted the researcher, a short interview regarding demographic and background information was conducted to see whether the participants fit the inclusion criteria of the study. Two persons who called to show their willingness to join the research were not included in the research because one of them felt very little guilt in his bereavement experience, and the other had lost siblings rather than parents. For valid participants the interview was usually arranged within 2 weeks according to the participants' convenience in time and location. The participants were given the choice to go to either the interview room in the Center on Behavioral Health (CBH) at the University of Hong Kong (HKU) or to chose another convenient or suitable place for the interview. In the end, two participants were interviewed at their homes, one participant was interviewed in a palliative care center and the rest were interviewed in the CBH. The process of recruiting new participants continued until data saturation was reached, in which no new theme in subjective experience of guilt comes out from the interview (Strauss & Corbin, 1998). After the data started to replicate existed themes, three additional participants were recruited and the interview with them was analyzed. This process confirmed that data saturation was reached because data in the last three participants did not bring new theme.

Demographic and death related information

In total 16 bereaved adult children were interviewed for their experience of guilt following the death of their parent(s). Eleven of them were women and five were men. They vary in age, bereavement time, relationship to the deceased, and in the cause of the related death; as planned in the sampling frame. Their demographic and other background information is presented in Table 4.2.

Table 4.2 Background Information of Participants

Code of Participants	Gender	Age	Marital Status	Religion	Deceased Parents	Bereavement Time	Cause of Death
F1	F	20	Single	Christian	Mother	5 months	Natural
F2	F	52	Single	Christian	Both	Father: 31 years Mother: 14 months	Father: Un – natural (Fall) Mother: Natural
F3	F	45	Single	Christian	Both	Father: 8 years Mother: 6 years	Both: Natural
F4	F	51	Married	Catholic	Father	2 years	Unnatural (Unknown)
F5	F	54	Single	Catholic	Both	Father: 30 years Mother: 1 year	Both: Natural
F6	F	54	Married	NA	Both	Father: 29 years Mother: 5 years	Both: Natural
F7	F	54	Married	Catholic	Both	Father: 13 years Mother: 6 months	Father: Natural Mother: Natural
F8	F	45	Married	Christian	Father	6 years	Natural
F9	F	48	Single	Catholic	Both	Father: 10 years Mother: 17 months	Both: Natural
F10	F	59	Married	Catholic	Both	Father: 43 years Mother: 11 months	Both: Natural
M1	M	24	Single	NA	Father	1 year	Un – natural (Accident)

to be continued

Code of Participants	Gender	Age	Marital Status	Religion	Deceased Parents	Bereavement Time	Cause of Death
F11	F	46	Single	NA	Both	10 years	Both: Natural
M2	M	42	Married	Christian	Both	10 years	Both: Natural
M3	M	45	Married	Christian	Father	6 years	Natural
M4	M	27	Single	No	Father	2 years	Natural
M5	M	28	Married	No	Mother	2 years	Natural

Procedure

The data were collected through face to face semi – structured interviews which lasted between 90 to 120 minutes. An interview schedule was prepared beforehand (see appendix 3). Both probing questions (to explore the main aspects of their guilt experience) and prompting questions were prepared, in order to invite the participants to further elaborate on their points or expressions when necessary.

At the beginning of the interview, a brief introduction about the research and the respondents' rights were introduced to participants. After they understood and signed the consent form, the interview began, with audio being recorded. The first part comprises self introduction and warm – up questions to help participants get used to the environment and feel relaxed enough to share their feelings. Then the interview continued, following the interview guide but not necessarily sticking to the initial sequence of the questions, depending on the flow of how the participants shared in their answers. In the end of each interview, I expressed thankfulness and appreciation to the participant and checked his/her emotional status. Before the start of the interview, all participants were informed that they can be provided with counseling and support by professional bereavement counselors if they feel stressed either during or after the interview. However, none of them made such a request.

Ethical consideration

Guilt is one of the most difficult feelings to release in bereavement, because most of the time, the bereaved persons' self – blame originates from irrational thoughts and feelings that to others may seem irrational, and thus they seldom express their feelings of guilt to other people, thinking that others will not understand. Meanwhile, the negative effect of self – blame on self worth also causes bereaved persons to be too ashamed to share their thoughts of self – blame and guilt (Miles & Demi, 1983). The interview, however, just provides a chance for those who are willing and ready to express their experiences, with an attentively listening by a professional and a nonjudgmental audience, in a safe atmosphere. Previous research shows that bereaved participants feel good participating in qualitative research, since this gives them a way in which they can express their experiences (Dyregrov, 2004). The autonomy of the participants was fully respected, and they all participated in the research on their own free will. Though this may present a risk to the present research of self selected sampling bias, this is unavoidable in bereavement research, in which the subjects' well – being is maintained as the highest priority. Additionally, participants were also free to withdraw from the research at any time if they felt too uncomfortable to continue.

All personal information from the participants was kept confidential. The documents and audio recordings containing contact and personal information were locked in a drawer to which only the researcher has access. All information referring to their identity was altered in the data analysis and research report. This data will be kept for up to 3 years following the project.

The researcher was alert to participants' emotions during the interview process. Informants were assured of their right to withdraw anytime without consequence. The Jessie and Thomas Tam Centre (JTTC), which provides bereavement counseling, and the Center on Behavioral Health at HKU were connected with the process to provide backup counseling support for those participants in need.

Ethical approval by the Human Research Ethics Committee for Non – Clinical Faculties in the University of Hong Kong was received before any data was collected.

Data analysis

Interpretative Phenomenological Analysis (IPA) was adopted as the data analysis method. IPA is a qualitative research method which originated within health psychology in the 1990s, to understand the experience of pain (J. A. Smith, 1996), since when it has developed within clinical and social psychology as well. It concerns primarily the subjective experiences of individuals and how they make sense of their own experiences. This method assumes that individuals are experts regarding their own experiences, and that they attempt to make psychological interpretations of their own experiences. It is acknowledged in IPA that the researcher may not objectively know another person's phenomenological world, but that they can probably get close to accessing it. The interpretative, rather than simply descriptive, nature of IPA has been recognized and embraced by researchers (Larkin, Watts, & Clifton, 2006).

IPA is a particularly relevant method in this study, as it puts bereaved adult children in the position of experts, to give voice to their own experiences and interpretations. IPA has been widely used to explore and understand various emotional experiences by researchers; for example, the experiences of anger (Eatough & Smith, 2006) and of boredom (Martin, Sadlo, & Stew, 2006). It is not used to test any causal hypoproject, but to shed insights on our understanding regarding participants' guilt experience, along with accurate narratives.

The data analysis adopts a systematic IPA procedure, as suggested by Smith

(2009) with the assistance of the ATLAS. ti 5.0 software. The first step is for the researcher to familiarize him/herself with the transcription of the interview from one participant, by reading through it again and again, and adding commentary on the transcription. The commentary can consist of any inspiring thoughts or even questions that the researcher asks to facilitate his or her interpretations of the test. The commentary is important to get the researcher to interact with the data, as well as to trace the data analysis audit trail. Then, preliminary themes are identified based on the text. Some themes have sub – themes under them to further elaborate different aspects of that theme, while other themes represent themselves well and do not contain sub – themes. After examining the connections between themes, they are grouped into super – ordinate themes, which are then given a more inclusive name. The data analysis then moves on to the next case, following similar procedures to the previous one. After all cases are analyzed, the common super – ordinate themes across cases are summarized and presented.

The rigor of data analysis was enhanced by following systematic data analysis procedures of IPA, as well as keeping audit trail along with process of collecting and analyzing the data. The insights and reflections generating from the face to face contact with participants were documented immediately after each interview. The thoughts and decisions made in coding themes were also recorded. The audit trail was helpful to remind myself that whether the result was grounded in the data.

Results

The analysis of the interview has revealed several aspects of the participants' phenomenological guilt experiences, as presented for an overview in Figure 4.1. However, the present research is only interested in the subjective experiences of bereaved adult children, for the purpose of conceptualizing this phenomenon. Therefore only findings regarding their subjective guilt experiences will be elaborated upon and discussed in this chapter. The remaining aspects are beyond the range of the present project, and due to the limitations of time and resources, they will be reported and discussed in other academic publications in the future.

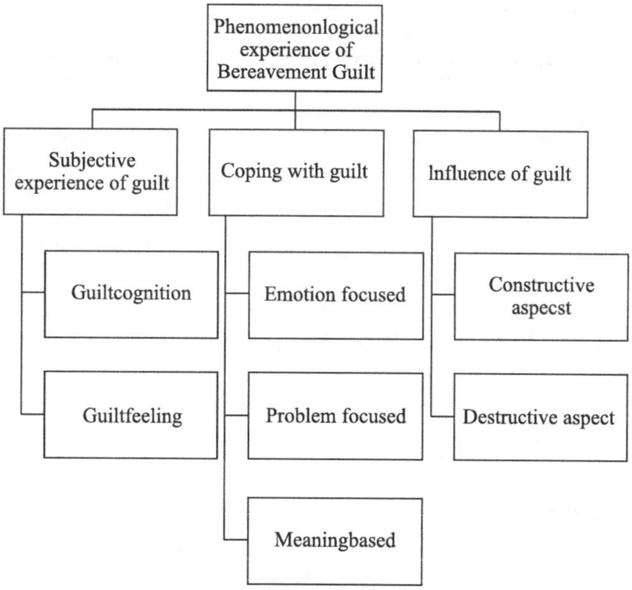

Figure 4.1 Aspects in Phenomenological Experience of Bereavement Guilt

The main objective of this study is to allow the bereaved adult children to define and describe their guilt experiences in order to provide material for conceptualization as well as expressions for use in developing the scale. The analysis of participants' subjective experiences of guilt revealed that both cognitive and affective dimensions exist within their subjective experience of guilt. The cognitive dimension concerns cognitive appraisals and evaluations which produce guilt, whereas the affective dimension illustrates their feelings associated with these guilt cognitions. The structure of these themes is presented in Figure 4.2.

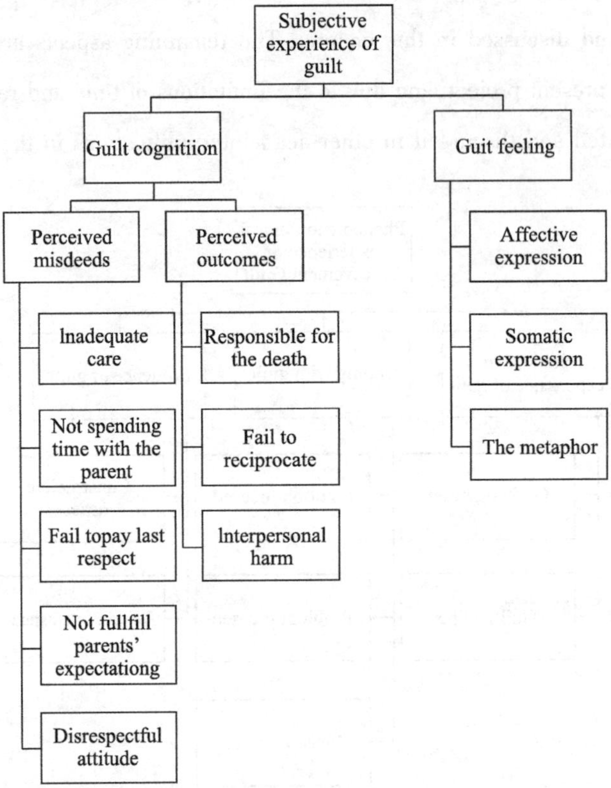

Figure 4.2 Themes of Guilt Experience in Bereavement

Guilt cognition

When asked about their guilt related experiences, each participant responded immediately with things about which they feel guilty. Their narratives comprise not only evaluations on past behaviors which failed to meet their inner expectations, but also on the unwanted consequences for which they held themselves responsible.

Perceived misdeeds

This theme illustrates their self – evaluation of past behaviors against inner standards, and of recognized discrepancies between actual behavior and inner expectations. The following sub – themes give an in – depth illustration of adult children's inner standards and expectations of themselves, including things they feel they should have done and things they should not have done.

Inadequate care

The adult children interviewed believed themselves responsible for taking care of and look after their aging parents. Therefore, the participants blamed themselves when they felt their parents did not receive adequate care. This care includes supporting their parents, providing them with necessary articles for daily life, being there when they need them, and protecting them from getting hurt.

F7's mother lived alone and F7 went to her mother's house now and then to take care of her, until the day her mother was diagnosed of cancer. After receiving treatment to control the decease, her mother went back home and stayed there. F7 considered hiring a domestic helper to live with her mother and look after her. However, she didn't feel it was very urgent so she did not put her plan into immediate action. One day her mother fell down at home and her health condition deteriorated from that point on. Whenever she thought about this after her mother passed away, F7 always blamed herself for her negligence and felt she did not do a good job looking after her mother.

"I didn't realize this was a problem until she fell. At first, it's just, you know, not necessary: (I thought) 'Wait for a while, wait for a while...' – till it happened, then I felt guilty. I think, 'Why, why did I treat it lightly?' I should have hired a care person for her. Even though no one could guarantee that she would not

fall down if there was a helper with her, at least there could have been someone to care for her daily life, someone who can help her, so she doesn't need to go out for grocery shopping, and the risk would be lower. This is the thing that makes me feel very uneasy." (*F7, woman, 54 years' old*).

F3's mother was living in an expensive elderly care center, but considering the economic pressure and her mother's insistence, F3 and her siblings found another, cheaper care center for her mother. The cheaper one did not provide as good support as the first one, and F3 felt her mother was not happy there. She expressed her guilt for not providing the best environment for her mother within her ability. One reason for this, so she said, was that she didn't realize that there was not much time left in her mother's life:

"*I think I could have done more in some aspects. At that time, we, at least I, didn't realize that her life would be so short. It's just... you don't think about it, that this old lady will die soon. It (the death) was sudden, and I feel I didn't fulfill my responsibility as a daughter quite well. I think, if, if I had known there would not be much time left in her life, I would insist upon her staying in the expensive health center. I should not have let her move to the new one.*" (*F3, woman, 45 years' old*).

When parents were in the hospital during their last days, their children would make every effort to provide them the best medical care. However, they just thought they "*should*" have done this, even though it may have exceeded their actual ability. For example, F1 was a 20 year old girl who had just started college, and had no income. She felt guilty for not being able to afford a specific medical treatment, which was said to be beneficial in treating her mother's type of cancer.

Furthermore, the frustration of not being able to provide better care or medical treatment for the parents makes them feel guilty and helpless. M4 describes his experience like this:

"*He got sick and got worse, but I could do nothing. As a son, there should be something I can do to help my father, but I can do nothing. I witness him getting feeble, then, I can't help him. I don't know (what to do), I am not a doctor. Even the Chinese medicine was expensive, and I had just graduated and started a job. I didn't have much money, and I could not afford the medicine. So I felt, how to say, helpless.*" (*M4, man, 27 years' old*)

It should also be noted that the types of care provided for parents include not only material and physical care, but also psychological and emotional care. F10 lost her sister from cancer one year before her mother passed away, and she felt guilty for not being aware of the influence of her sister's death on her mom. Even though she had done a good job in satisfying her mother's material needs, she still blames herself for not providing adequate emotional support to her mother.

"*One of the most important things to her was her feelings. I didn't show much concern for her inner feelings. On the surface, I have provided a lot of things for her. I have totally, I mean materially, satisfied her needs. I have no guilt regarding those things. But I feel guilty about (not taking care of) her heart feelings, very guilty. How to take care of her feelings, to make her feel really happy from her inner heart? I feel guilty: I did not do enough, I really did not do enough.*" (*F10, woman, 59 years' old*)

Not spending time with the parents

Nearly all participants felt guilty for not spending as much time as they "*should*" with their parents.

Adult children are in the development stage of leaving their parents and building their own social networks and core families. They are busy with work or with their own families, until the parents have passed away; which is like a wake up call to make them aware that they had not spent as much time with their parents as they had wished, and now they won't have the chance to do it anymore.

"*I feel I did not spend enough time with him, and this feeling emerged after he left this world. I was busy with my work and my own family every day. I had just got married then, so I spent little time getting along with him. I was not able to visit him in the hospital even during his last several months. Sometimes, he was allowed to stay at home as well, but I still could not go home to see him every day. So when he left, I felt unhappy and sad in my heart, because I really did not devote much time to him.*" (*M3, man, 45 years' old*)

M6 was studying in another city when his mother was diagnosed with cancer. He went back to visit his mother several times, but did not stay there with her because he had to go back to the University to study. After his mother passed away later, he felt very guilty for not "*asking for a leave or even suspending my study for a year to stay together with her*".

Paying last respects at the time of death

One informant mentioned her failure to be with her parents at the moment of the death to show her last respects. She was at work that day, and when she got the notice from the nurse and rushed to the hospital, her mother had already passed away. She felt it is the responsibility of Chinese children to pay their last respects to their parents at the moment of death, however she failed to do that. She expressed her guilt like this:

"*She was leaving us to begin a long journey far away, but I was not there to say goodbye to her at the last moment. Even though I had said goodbye to her several times before that, but still, she left in loneliness, all by herself. We Chinese need our children to be there at the last moment. It's so important, but I could not do it. I couldn't do it for my father, nor for my mother." (F2, woman, 52 years' old)*

Not fulfilling parental expectations

Some of them also felt guilty when they feel their parents had invested a lot in them and held high expectations for their achievement; and yet they had not accomplished achievements big enough to reflect such expectations.

"My guilt is connected to feeling indebted to him. He gave me a family, raised me up, and sent me to a famous school, but I did not study hard to achieve. If I had studied hard, I could have improved our family (economic) conditions, and I could have fulfilled my parents' wishes. I know he wished me to have some achievements, like expecting me to become a doctor or a lawyer, but I did not achieve that. It's about answering his expectations, you know, not giving back in material." (M2, man, 42 years' old)

Disrespectful attitudes

During the growing up process, there must be times when conflicts occur between parents and child. When the child recalls those incidents along with their bad attitudes displayed in those conflicts and episodes, they feel guilty for not behaving appropriately, and feel sorry for their disrespectful attitude toward their parent(s).

"I feel guilty because I really did something wrong: for example, I offended him. Sometimes when I think back, I recall those times when I said inappropriate words to him, I ran away from home, and I had conflicts and quarrels with my father. I do dwell on those fragments." (M3, man, 43 years' old)

Perceived consequences

In addition to the narratives about misdeeds that made them feel guilt, the adult children shared more cognitive content as well, regarding the unwanted consequences for which they believed their behavior was responsible. This theme is helpful in understanding why their behavioral failure to meet inner standards and expectations is significant enough to induce guilt. The reason for this can be further illustrated by the perceived consequences of those actions or inactions. In other words, the adult children believe their failure related to inner standards and expectations contributes to unwanted consequences for them, including the death of their parents, not being able to reciprocate with the parents, and even harmful effects imposed on the parents.

Responsible for the death

Bereaved people usually think about the causes and potential preventions of the death. It is common for these bereaved children to feel that their actions or inactions have played a role in their parents' death. In some cases, adult children felt that they had directly contributed to their parents' death, even though they know the detrimental cause was actually disease or other external reasons. For example, F4 had a lot of conflict with her father, especially after her father had experienced a stroke. This situation lasted for 4 years, and one day the father died suddenly after falling down on the floor. F4 felt as if she had made her father so mad that she was the reason for her father's death.

"I used to have a lot of fights with my father. It felt like I could do nothing right in his eyes. He used to scold me with mean words. Now when I think back, it was because his health was getting worse. Then he was admitted to the hospital. He developed a fiery temper, and often beat me. So I talked back to him and fought

back. One time he got so angry that he nearly opened the window to jump… After he died, I felt worthless; I made my father so angry. I feel as if he died from his angry toward me." (*F4, woman, 51 years' old*)

In more cases, the adult children looked back and searched in the past for things they could have done to prevent the death at that time. Even though they did not contribute to the death directly, they still felt responsible for the death, since they believed it could have been prevented had they done things differently. The role they think they could have played to change the events are addressed in the following appraisals.

"*It was because I did not stay together with her, so she was careless about her health. If I was there with her, it (the cancer) would have been detected earlier in her regular check ups, and it wouldn't have developed to such a serious degree.*" (*M5, man, 28 years' old*)

"*As a nurse, why didn't I detect any sign of it (the cancer). If I could have detected it earlier, then some treatment could have been done, and maybe she would not have died, or at least not at that time.*" (*F11, woman, 46 years' old*)

"*The relationship between my mother and my father was not good, and they fought all the time. I think maybe mom got cancer because she was unhappy. Why didn't I step forward to stop their fighting at that time?*" (*F1, woman, 20 years' old*)

Failure to Reciprocate

There are things adult children believe they should have done for their parents, because their parents have given them so much, and they feel obligated to pay their parents back. Therefore, their guilt can also arise from how much their parents have given to them, compared how little they have paid back in return.

This guilt stems from the in – equality in the relationship between giving and taking, as well as from their awareness that the chance to pay back their parents has been taken away by the death.

"I think she has sacrificed so much for us, but I can do so little for her. It feels like I can never pay her back." (*F1, woman, 20 years' old*)

"If you knew what my father was like during my growing up years, you would know that he is a perfect father, who deserves 100 points [on the father scale]. But at the same time, I only deserve 50 points as a son. So I think compared with his giving (to me), I have given him only a small amount." (*M3, man, 45 years' old*)

The more parents have contributed to them or sacrificed for them, the more they feel obligated to reciprocate; through bringing good things to their parents, taking care of their parents, and fulfilling their parents' expectations. However, the death makes it impossible for them to give back to their parents, and induces their sense of guilt.

"I wanted to take care of him. In my family, my parents have invested a lot to take care of us. My parents don't go out to play. They put all their time, money, and energy toward us. However, he left this world when I finally became independent and capable of taking care of him. So I feel guilty because I cannot take care of him." (*M3, man, 45 years' old*)

Hurting the parents

Another source of adult children's guilt arises when they have imposed harmful effects on their parents. The children blame themselves for causing harm to their parents, hurting the parents or making the parents upset.

"For example, I had quarrel with him and I ran away from home. To tell the truth, this was hurtful to my parents. No matter how many times I told him I was

sorry, the hurt was already done, and it did upset them. So I think I did wrong, and I remember this even though they didn't take it to heart." (*M3, man*, 45 *years' old*)

The worse the consequences were, the more guilt they felt. When they sympathize with their parents pain which they believed their own misdeeds to have caused, their feelings of guilt become even more intensive. For example, F2 recalled an incidence in which she was irritated by her father's nagging, so she talked back angrily to her father. She believed her response hurt her father's feelings, and she felt very remorseful and guilty for having such a bad attitude.

"*I got irritated, so I reacted to him in a very bad attitude. Now he is gone, and I realize how selfish I was. I was only concern about my own daughter and grandchild, and I did not understand his concern for me. I even told him things like: 'Just leave me alone!'*" (*F4, woman*, 51 *years' old*)

Her guilt became more prominent when one day her daughter talked back to her angrily, just like similar scenarios that had happened between her and her father. At that moment, her empathy towards her father's distress, which she believed was imposed by her misdeeds, made her feel even guiltier.

"*He must have felt pained when I talked back to him, you know? Just like when my daughter talked back to me my heart felt hurt. I can perceive his feelings through experiencing my own feelings. When my daughter talks back to me, I feel my pain; so I feel even guiltier (regarding my father)*…" (*sobbing*) (*F4, woman*, 51 *years' old*)

The Feelings

It may be difficult to ask people about their feelings when they feel guilt, because guilt is already a word used in daily life to express a particular emotion. However, participants tried to use different ways to describe their feelings of guilt.

The expressions they used to describe this feeling provides vivid illustrations of the affective features of guilt in bereavement. People used different kinds of expressions to describe their feelings, including emotional, somatic and metaphorical methods. In general, guilt in bereavement is perceived as unpleasant and distressful.

Emotional expression

Guilt does not feel good. It is an unpleasant emotional experience, accompanied by the sense of being blamed, but the intensity of the distress varies across individuals.

In the intense cases, guilt is "painful":

"I blame myself, yes, it's just like you are blaming somebody else, and are blamed by somebody else. It's a very unpleasant feeling…If you have to analyze it, it's pain in your heart. Yes, it's painful." (*M2, man, 42 years' old*)

In some cases, the feeling was not so strong, and words such as "unsettling" "unhappy" or "sad" were used to describe the feelings.

Somatic expression

Another way to express their feeling in guilt is through describing their somatic sensation. One participant said she felt headaches whenever she recalled things that she felt guilty for, while others described the feeling as related to their heart. For example, F6 said she felt a "tense" and "sour feeling" in her heart, while M2 said;

"(*It was*) *Very hurtful, miserable, nothing but pain. It's not pain in your body, it's a convulsion in your heart.*" (*F6, woman, 54 years' old*)

Similarly, F4 also used words to express the feeling as being in her heart:

"*My heart feels pain, as if it were twisted.*" (*F4, woman, 51 years' old*)

It is interesting to notice that guilt is referred to as a "conscience emotion" as it is a feeling arising from the conscience awareness of violating inner standards, while the literal translation of "conscience" in Chinese is "good heart". When people feel guilt, their unpleasant somatic expressions are mainly focused on their heart, as if their "good heart" got hurt.

The metaphor

Some people used metaphors to illustrate their feelings. These metaphors concern negative connotations. For example, the color of the feeling may be "black" or "grey", or it may be "heavy", "oppressing" and may cover the person like a "dark cloud".

"(*It's*) *Very heavy, oppressing your heart. It feels black and like a huge pile whenever you think about it.*" (*F2, woman*, 51 *years' old*)

"*It feels grey, just a sense of grey, like a dark cloud on top of your head.*" (*M4, man*, 27 *years' old*)

"*It feels like pressure on your body. I felt I was squeezed into a little room, very distressful.*" (*F1, woman*, 20 *years' old*)

Meanwhile, the self – blame in their guilt was like a weapon used for self attack: they were hurt and felt pain. F10 described her feelings like this:

"*When you think about it, it's like, at first, it's was very severe, as if you keep stabbing and thrusting yourself with a knife. It was really miserable.* (*F10, woman*, 59 *years' old*)

Discussion

The results of the interview provide rich information about bereaved adult children's bereavement guilt experiences, which comprise both cognitive and affective dimensions.

Guilt cognitions

Appraisal on both behaviors and consequences

When talking about their guilt in bereavement, the first reaction of participants was related to things about which they feel guilt, which implies the importance of guilt cognition in bereaved adult children's subjective guilt experience. One aspect of their cognition is related to specific behaviors which they evaluated to be failure related to their inner standards or expectations, as included in the definition of bereavement guilt. Moreover, the qualitative findings have suggested another important aspect of their cognition; namely, that related to the perceived consequences of their behavioral failure. This illustrates that bereaved adult children's guilt over their past actions or inactions is due to the consequences perceived to be contributed by those behavioral failures.

Responsibility for the death

Guilt regarding responsibility for the death is a unique feature of guilt in the domain of trauma and loss, and is different from guilt in other contexts. In fact, this has been the main focus of guilt in previous bereavement literature (Downey et al., 1990; Horowitz et al., 1984; Weinberg, 1994). Self attribution of the loss has been suggested to be one way to provide an answer for those who have encountered an unwanted event (e.g. trauma and loss), related to the question: "Why did this happen?" Self attribution for the loss can help the individual to preserve a sense of control in order to combat the sense of helplessness in facing a death or traumatic event, even at the cost of self-blame and guilt (Janoff-Bulman, 1992).

Harmful effects toward the parents

This theme is consistent with the moral feature of guilt, which is felt when one perceives his or her behavior as being responsible for another's distress (Kugler & Jones, 1992). Participants said they felt guilty because their behavior has hurt their parents or made their parents unhappy. As shared in the interviews, the more empathy they felt toward the parents' distress, the more guilt they felt. Moreover, this kind of guilt is suggested to have a bearing on the interpersonal functions that re - distribute distress in the relationship and maintain emotional equality (Baumeister et al., 1994). The distress of guilt indicates bereaved adult children's acknowledgement of their transgression, and also shows how much they care for their parents and for the relationship.

Failure to reciprocate

It is important to notice that failure to reciprocate is also one of the main components of Chinese adult children's guilt, yet this has not been stressed in western literature on guilt. This is probably due to the culture differences in moral and social standards between eastern and western countries. Western society emphasizes independence and protects individual rights. Therefore their moral system is built on the principle of protecting individuals from getting hurt. In other words, not hurting others is a crucial moral code in their system. However, eastern society emphasizes relationships and societal harmony, and therefore the moral system stresses not only negative duties (do no harm to others), but also positive duties (obligations within the relationship) (Bedford & Hwang, 2003). Reciprocation is one of those positive duties that has been considered to be an important social rule by which to regulate the interpersonal interactions in Chinese society (Hwang, 1987). It has also been identified by researchers as a crucial attribute in filial piety, which is a traditional virtue and has significant implications within inter – generational relationships in Chinese culture (Yeh, 2002). Considering the parents' giving, and the perceived responsibility to pay back their parents, this has been found to motivate Chinese adult children to take care of their aging parents (Wong & Chau, 2006). Though filial responsibility may be a universal construct (Gans & Silverstein, 2006), the theme of reciprocation to parents is more prevalent in Chinese than in Western countries (Rothbaum & Xu, 1995). While reciprocation is a filial obligation to parents, failure to reciprocate induces guilt on bereaved adult children.

The feelings of guilt

The affective component of guilt is also revealed through vivid emotional, somatic and metaphoric expressions shared by these adult children. All their expressions indicate that guilt is an unpleasant and distressful feeling, but do not suggest different clusters within the affective component. The intensity of the distress of guilt varies across individuals, and can range from extremely uncomfortable to painful (in severe situations). The pain of guilt should be distinct from the pain of grief, because the participants were describing the feelings accompanying their guilt cognitions, and described it as a self – blame related pain.

The multidimensional conceptualization of guilt

Previous conceptualizations of guilt in bereavement are composed of only the cognitive aspect, whereas the affective component was missing (Miles & Demi, 1983). However, as revealed in the qualitative findings, guilt includes not only cognitive appraisals but also distressful and unpleasant feelings. Moreover, the content of guilt cognition is more than self attribution for the death, which was nearly the only focus of previous bereavement researchers' investigations on this topic. The cognition shared by participants was found to be related to different kinds of behavior, and two kinds of additional consequences beyond responsibility for the death were mentioned as well. Those consequences carry with them interpersonal and cultural features, as mentioned previously, and should not be ignored when investigating guilt among bereaved people. Therefore, it is supposed that bereavement guilt is a multidimensional construct, which comprises both cognitive and affective components, and that there are various elements within guilt cognition. Meanwhile, based on the qualitative findings, bereavement guilt is re - defined as:

A remorseful emotional reaction within grieving, accompanied by the recognition of having failed to live up to one's own inner standards and expectations in the relationship to the deceased, and of being responsible for unwanted consequences.

The missing component, survivor guilt

It is also noticed that survivor guilt, which has been reported by bereaved parents in previous studies (Miles & Demi, 1991, 1992), was not reported by participants in this study. This might due to the different nature of parental loss and adult children's loss. Death of a child for parents is usually an unexpected event and violates common expectations and natural laws. Therefore, parents who have lost a child are more inclined to feel they should have died instead of their child (survivor guilt). In contrast, most people lose their parents as adults, and this does not violate natural order and social expectations. Therefore, bereaved adult children may have less risk of survivor guilt. However, the present study does not mean to generalize the findings based on such a small number of participants, to the whole population. It is possible that those guilt cognitions do exist in bereaved Chinese adult children, but due to low incidence (Miles & Demi, 1991, 1992), it was simply not identified in the present study.

Implications for the next study

The interview has provided vivid narratives for understanding bereaved adult children's guilt experiences. It identifies the affective feature of guilt, as well as the interpersonal and cultural characteristics of the phenomenon. It suggests that bereavement guilt is a multidimensional construct. The structure of this construct will then be built up through exploration in study two. The narratives collected in this study will in the next chapter be translated into items for developing a bereavement guilt scale, and for exploring the structure of bereavement guilt.

Limitations

The present study has focused only on a specific subgroup of bereaved people. Some themes may bear special meaning in this subgroup (e. g. paying the last respects to dying parents), due to the distinct features of the parents – child relationship within this subgroup. Meanwhile, all findings derive from narratives taken from only a small group of participants. It cannot and does not aim to generalize the findings to represent the general population. Therefore, more large scale studies are needed to test whether the findings in this study are transferable to other subgroups or to general populations.

Chapter 5

Development and validation of a Bereavement Guilt Scale

Introduction

In order to explore the relationship between guilt and health outcomes in bereavement, it is necessary to have a valid measurement to assess bereavement guilt. Researchers have previously advocated developing a domain specific measurement for guilt, acknowledging the importance of the context (Tilghman – Osborne et al., 2010). For example, there is the care giver guilt questionnaire (Losada et al., 2010), and the trauma related guilt scale (Kubany & Abueg, 1996), developed by researchers in their respective domains. However, there is no valid bereavement guilt scale related to the area of death and loss. As stated in the literature review, the measurements used by previous bereavement studies lack psychometric validity and fail to capture the complexity and multidimensional feature of guilt in bereavement. Therefore, a valid instrument by which to fill in those gaps is important and necessary to gain more understanding on this concept; such as its prevalence, risk factors and relationship to maladaptive grief adjustment.

The conceptualization of bereavement guilt has been constructed from the literature review and qualitative findings. Bereavement guilt contains both cognitive and affective aspects, and there are different components within guilt cognition. In this study, a scale will be developed that is based on such conceptualizations and on the collected qualitative data. Its validity and reliability will also be assessed to ensure its psychometric properties. Using this scale, the conceptualization and structure of bereavement guilt will be further confirmed by quantitative data.

Objectives:

The present study aims to:

1) develop and valid the Bereavement Guilt Scale(BGS), and to

2) explore and confirm the structure of bereavement guilt.

Specifically, it will answer two questions:

What is the factor structure of bereavement guilt?

What are the psychometric properties of the BGS ?

Method

The development follows procedures suggested by De Vellis(2003), and is validated by quantitative data collected from an on – line survey among bereaved Chinese adult children. An online survey is a convenient mean of reaching potential participants in bereavement study, and has been widely used in recent years (eg. Boelen & Lensvelt – Mulders, 2005). It is found to attract a higher response rate in some studies than mailed questions(Evans & Mathur, 2005).

Developing the scale

The item pool was generated both from the qualitative study and from reviewing previous related literature. Since in the conceptualization, guilt comprises both cognitions and feelings, participants' narratives responding to the two questions: "What do you feel guilty about?" and "Can you describe to me your feelings when you felt guilty?" were selected from the transcriptions of the interview in study one. After that, interviewees' narratives were translated into sentences and expressions for the items pool. This process identified 64 guilt experience expressions. I scrutinized these sentences and delete duplicate ones, after which 25 items were generated to capture the content of guilt. At the same time, after reviewing previous literature about guilt in bereavement, or grief scales that contain content related to guilt (Boelen & Lensvelt - Mulders, 2005; Hogan, Greenfield, & Schmidt, 2001; Sanders et al., 1985), some guilt expressions that were documented in the literature but not reflected from the interview were added into the items pool in order to maximize the comprehensiveness of the newly developed scale. Those expressions were mainly about survivor guilt; i. e. guilt about continuing to live and enjoy life after the loved one's death (eg. "I should not live since he is dead."). Therefore, the final item pool contains 29 items. Those items were discussed by an expert panel, which included one (1) professor in the bereavement research area, three (3) PhD students specializing in the area of death and dying, one (1) social worker and two (2) counselors. The face validity and content validity were guaranteed through this process of expert review (De Vellis, 2003). Expression modifications were made to ensure clarity, following the discussion. This preliminary version of the bereavement guilt scale contains 2 sub - scales: one about guilt cognitions, which reflects the various content of people's guilt; and the other about guilt feelings, which captures the affective experience of guilt.

Procedure

The recruiting advertisement and introduction to the study was disseminated through two Chinese on - line memorial websites[1]. Users of those two websites are all bereaved people. They can build up memorial pages for their deceased family members on these websites, and they can also get involved with other mourning activities on - line, such as creating a virtual grave yard, write a life review book for the deceased person, or write letters to the deceased. There are also on - line forums for those website users, as well as an information page which disseminates news, education materials and information for grieving people. The recruiting

---

[1] They are zupulu (http://www.zupulu.com/) and Waheaven (http://www.waheaven.com/)

information and link to the on – line questionnaire survey was posted on the first page of these two websites. After clicking on the survey link, anyone interested in the study was directed to an introduction page about the study objectives and the inclusion criteria for target participants. It explained that this survey contains a series of questions about their grief experience, and would take them about 20 ~ 30 minutes to complete. The confidentiality of their information and their right to withdraw at any time was also stated in the introduction (see appendix 4). As a token of appreciation, after they finished all the questions, they received a certain amount of virtual coins provided by the participating websites. The introduction also stated that qualified participants were Chinese adults who had lost their first degree relatives (including parents, spouses, children, and siblings)[1] and for whom the death had transpired within the preceding 10 years. If they fit the criteria, and were willing to join the study, they could state their agreement by clicking the "agree" button to enter the survey page (Appendix 5). For those who clicked the "do not agree" button, the page would close automatically.

Participants were invited to leave their on – line contact information in the last page of the survey, because a certain number of virtual coins provided by the two on – line memorial websites would be given to valid participants.

Ethical approval was sought and granted by the Human Research Ethics Committee for Non – Clinical Faculties, University of Hong Kong, prior to data collection.

Participants

There were in total 933 bereaved adult children who completed all the ques-

[1] Because this is one part of a big research project about bereaved Chinese people, the data collection was not restricted to bereaved adult children. However, only the data from bereaved adult children was selected and included in the analysis in this thesis.

tionnaires. The data was carefully checked for suspicious patterns, for instance, where same person fill out the questionnaire more than once. This was scrutinized through demographic variables and the contact information left by the participants. Some data did not fit the inclusion criteria, such as participants younger than 18, or those who had lost their parents more than 10 years prior to the survey. Those data were deleted from the final data set. This purging process removed 60 cases from the analysis, making the final sample size 873. The recommended sample size for conducting reliable EFA is preferably 10 times the number of items in the scale (A. B. Costello & Osborne, 2005). Considering that there are 26 items about guilt cognition to be analyzed through EFA, the minimum sample size for EFA is 260 participants. Therefore, this sample was split into 2 groups via random selection. One group comprises 273 individuals, and will be analyzed with Exploratory Factor Analysis (EFA) and item analysis, while the remaining 600 individuals' data will be analyzed through Confirmatory Factor Analysis (CFA).

A small sample was selected from the total to check for test – retest reliability. An invitation to fill in the Bereavement Guilt Scale again was resent to 50 participants, who were selected randomly, 3 weeks after they first participated. Twenty – four of them replied to this request and fill out the scale again 3 to 5 weeks after the first time they filled out the questionnaire.

Participants in this study aged between 18 to 72 years ($M = 42.21$; $SD = 10.25$), and lost their parents between 1 to 117 months prior ($M = 26.15$; $SD = 25.85$). Other demographic and death related characteristics are presented in Table 5.1.

Measurements

Demographic information and other death related information, such as period of bereavement, cause of the death, and psychological preparation for the death, were embedded in the questionnaire.

Guilt: The Bereavement Guilt Scale (BGS) contains 29 items, with 26 items reflecting guilt cognition and 3 items about guilt feelings. On each item, the participants were asked to indicate the degree to which the item describes their actual experience on a 5 point scale (1 = does not describe very well, 2 = does not quite describe me, 3 = describes me fairly well, 4 = describes me well, 5 = describes me very well).

Self esteem: The Rosenberg Self – Esteem Scale is a well validated 10 – item scale, which measures global personal self – esteem (Rosenberg, 1965). The Cronbach alpha of this scale in the present sample was 0.802.

Self – forgiveness: One question was asked to assess the degree to which participants forgive themselves for things they did wrong and feel guilt for: "To what degree have you forgiven yourself for things that made you feel guilty during bereavement?" Participants can respond with a choice from 0 (not at all) to 10 (completely). This question has good face validity, and is used as a validity check item in other studies (Wohl & DeShea, 2008).

Depression and Anxiety: Participants' mental health was evaluated by the Hospital Anxiety and Depression Scale (HADS) (Leung, Ho, Kan, Hung, & Chen, 1993), which contains 7 items to measure anxiety and 7 items to measure depression. It has been widely used among different populations and has proven to perform well in assessing symptom severity across various populations (Bjelland, Dahi, Haug, & Neckelmann, 2002). The Cronbach alphas in the present sample are 0.726 and 0.837 for anxiety and depression subscales respectively.

Anticipation of the death: Participants stated their level of psychological preparation for the death through answering the question: "Did you have any psychological preparation for the death before your parents passed away?" They selected from among 4 alternatives (1 = Not at all; 2 = A little bit; 3 = Quite prepared;4 = Totally prepared).

Table 5.1 Demographic and Death Related Characteristics

	N	Percent(%)
Gender		
Male	435	49.8
Female	438	50.2
Age		
18 ~ 30	128	14.7
31 ~ 40	253	29.0
41 ~ 50	317	36.3
51 ~ 60	148	17.0
60 ~ 72	27	3.1
Education		
Primary school and below	5	.6
Middle school	188	21.5
Adjunct college	328	37.6
Bachelor	292	33.4
Graduate school and above	60	6.9
Religion		
No religion	691	79.2
Buddhism	142	16.3
Taoism	10	1.1
Christian	21	2.4
Other religion	9	1.0
Marital status		
Single	84	9.6
Married	742	85.0
Divorced	41	4.7
Bereaved	6	.7
Time since death of the parents		
Within half a year	204	23.4

to be continued

	N	Percent(%)
Half to 1 year	166	19.0
1 to 2 years	186	21.3
2 to 5 years	192	22.0
5 to 10 years	124	14.2
Cause of the death		
Chronic illness	461	52.8
Sudden illness	313	35.9
Accident	58	6.6
Suicide	15	1.7
Others	5	.6
Unknown	21	2.4

Note. Time since the death was unknown for one participant.

Data analysis plan

The dimensionality of the BGS will first be explored by means of EFA and then tested by CFA. Items in the BGS will be analyzed from their loadings on each factor, communality, inter – item correlations and item – total correlation. The construct validity of the BGS will be assessed by the associations between BGS scores and constructs which are theoretically proved to be correlated with guilt, including self – forgiveness, self – esteem, depression and anxiety.

When people feel guilt and blame themselves, it is likely that they do not forgive themselves for their transgressions or failures. Therefore, it is supposed that guilt is negatively correlated with self – forgiveness (Strelan, 2007; Zechmeister & Romero, 2002). The feelings of guilt and negative evaluations of one's own behavior will also associate with negative evaluation of oneself (Boelen & Lensvelt – Mulders, 2005), and impair an individual's self – esteem (Tangney et al., 1995).

Meanwhile, researchers, as well as previous research findings, have established a positive correlation between guilt and depression and anxiety (Harder et al., 1992; Harder & Zalma, 1990). Therefore, the correlation between BGS scores and self – forgiveness, self – esteem, depression and anxiety will be assessed to provide evidence for the construct validity of the BGS.

Finally, the reliability of the BGS will be examined from its internal consistency and test – retest correlation.

Results

Explanatory Factor Analysis (EFA)

At first, all items from the BGS were subjected to EFA. However, the items describing guilt cognition and guilt feelings did not load on separate factors as was predicted in the conceptualization of bereavement guilt. This may be due to the potentially high correlation between guilt cognition and guilt feelings. Re – investigation of the present conceptualization of bereavement guilt reveals that only the structure of guilt cognition was subjected to exploration, because it was expected to be a multidimensional structure. At the same time, guilt feelings is conceptualized as being unidirectional, with a higher score indicating higher distressful feelings. Therefore, items on guilt feelings should not be included into EFA accompanying items on guilt cognition. A similar practice of combining both theory driven and data driven methods when developing scales has been documented in previous literature. For example, in the development and validation of the Trauma – related guilt scale, EFA was only applied to items regarding guilt cognition, while items about general guilt feelings were excluded from EFA (Kubany & Abueg, 1996).

Therefore, EFA was only applied to the 26 guilt cognition items using principal – axis factoring in SPSS with the 273 randomly selected participants from the total sample. The principal – axis factoring method was preferred over principle factor analysis because it is suggested to be more suitable for identifying latent variables, while the latter is mainly adopted as a data reduction method (A. B. Costello & Osborne, 2005). The items were subjected to direct oblimin rotation because it is possible that some factors are correlated. The scree test suggested by Cattel (1966), the eigenvalues – greater – than one rule and the interpretability of factors were all taken into consideration in determining the number of factors to retain.

The factorability of the correlation matrix and also the sampling adequacy should be assessed to determine whether factor analysis is appropriate to be applied to these items. These prerequisites are tested by using Bartlett's test of sphericity, which is used to prove that the correlations within a matrix are not 0, and by the Kaiser – Meyer – Olkin (KMO) measure, which measures the relationship between partial correlations and the sum of squared correlations. The results of these tests suggested that the data is suitable for factor analysis, as the Bartlett's test of sphericity is significant and the KMO value is 0.937 (See Table 5.2).

Table 5.2 KMO and Bartlett's Test

Kaiser – Meyer – Olkin Measure of Sampling Adequacy.		.94
Bartlett's Test of Sphericity	Approx. Chi – Square	4396.76 **
	df	325

Note. ** : $p < .01$.

The first round of EFA suggested 4 factors on the basis of the above criteria (4 component's eigenvalues are greater than 1, and the scree plot is presented in Figure 5.1). Those factors accounted for 55.83% of the total variance in all items (Table 5.3). The communality and factor loadings of each item are presented in Table 5.4 and Table 5.5 respectively.

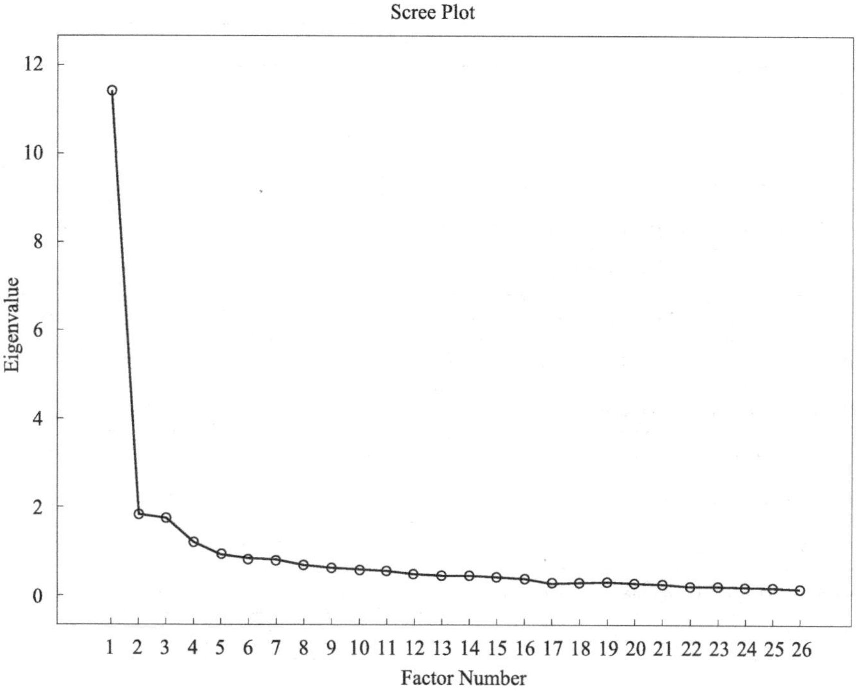

Figure 5.1 Scree Plot of Initial EFA

Table 5.3 Total Variance Explained by Initial Four Factors

Factor	Initial Eigenvalues			Extraction Sums of Squared Loadings		
	Total	% of Variance	Cumulative %	Total	% of Variance	Cumulative %
1	11.409	43.882	43.882	11.001	42.310	42.310
2	1.842	7.085	50.966	1.378	5.299	47.608
3	1.754	6.748	57.714	1.325	5.096	52.705
4	1.220	4.694	62.408	.813	3.128	55.833

Table 5.4 Communalities of Initial EFA

	Initial	Extraction
BGS1 I did not provide the best care to **	.586	.523
BGS2 I treated ** with a bad attitude	.624	.569

to be continued

	Initial	Extraction
BGS3 I was not considerate enough to **	.679	.636
BGS4 I think I should not be happy anymore since ** has passed away	.437	.461
BGS6 I feel guilty for my own survival following his/her death	.569	.662
BGS7 I did not try everything I could to improve our relationship	.469	.448
BGS8 There are many things I did not do for **	.608	.607
BGS9 I blame myself for not being with ** at his/her last moments	.324	.256
BGS10 I did something hurtful to him/her	.636	.562
BGS11 I failed in my filial piety toward him/her	.748	.752
BGS12 I successfully fulfilled my duties as his/her child(reverse scored)	.447	.317
BGS13 I feel ** gave me a lot of good things, yet I could not reciprocate	.495	.529
BGS14 I made ** unhappy	.618	.540
BGS15 God is punishing me by taking away his/her life	.579	.497
BGS16 I feel responsible for his/her death	.695	.710
BGS17 He/she suffered because of me	.702	.699
BGS18 I feel sorry toward ** whenever I am happy after he/she died	.527	.549
BGS20 I did not care for ** enough	.659	.662
BGS22 He/She might still be alive if I had done a better job	.631	.639
BGS23 I failed to give ** good things for his or her welfare	.695	.702
BGS24 I think his/her death is retribution for my wrong doings	.607	.605
BGS25 I think he/she would not have died at that time if I had done things differently	.606	.614
BGS26 I could not reduce or share his/her pain	.329	.259
BGS27 I failed to achieve his/her expectations toward me	.567	.509
BGS28 I feel indebted to **	.666	.659
BGS29 I did not spend enough time with **	.512	.548

Note. ** represents the deceased family member.

Table 5.5 Rotated Factor Loadings for Initial 26 Items of Guilt Cognitions

	Factor			
	1	2	3	4
BGS29 I did not spend enough time with **	.634			
BGS13 I feel ** gave me a lot of good things, yet I could not reciprocate	.633			
BGS8 There are many things I did not do for **	.588			
BGS1 I did not provide the best care to **	.482			
BGS23 I failed to give ** good things for his or her welfare	.466			-.415
BGS9 I blame myself for not being with ** at his/her last moments	.459			
BGS28 I feel indebted to **	.350			-.327
BGS26 I could not reduce or share his/her pain	.300			
BGS6 I feel guilty for my own survival following his/her death		.751		
BGS18 * * I feel sorry toward ** whenever I am happy after he/she died		.652		
BGS4 I think I should not be happy anymore since ** has passed away		.619		
BGS2 I treated ** with a bad attitude			-.804	
BGS3 I was not considerate enough to **			-.776	
BGS14 I made ** unhappy			-.688	
BGS10 I did something hurtful to him/her			-.682	
BGS11 I failed in my filial piety toward him/her	.398		-.617	
BGS7 I did not try everything I could to improve our relationship			-.611	
BGS20 I did not care for ** enough	.443		-.445	
BGS12 I successfully fulfilled my duties as his/her child(reverse scored)			-.417	
BGS27 I failed to achieve his/her expectations toward me	.312		-.369	
BGS16 I feel responsible for his/her death				-.751
BGS25 I think he/she would not have died at that time if I had done things differently				-.727
BGS22 He/She might still be alive if I had done a better job				-.696
BGS17 He/she suffered because of me				-.610
BGS24 I think his/her death is retribution for my wrong doings				-.532
BGS15 God is punishing me by taking away his/her life		.337		-.437

Note. Factor loadings of less than 0.4 are not displayed in the table.

** represents the deceased family member.

With these statistical results, items were analyzed to see if they: 1) failed to load on either factor with a factor loading greater than 0.4; 2) have cross – loadings above 0.4, and the difference between two loadings is less than 0.2; and 3) have communality extraction less than 0.4. The item will be removed if it fits any one of the above criteria. The analysis process resulted in 11 items being removed from the scale. After removing those items, similar methods were used in conducting the EFA again. The Bartlett's test and KMO test on the remaining 15 items suggested suitability for EFA, as presented in Table 5.6. The final EFA still suggested 4 factors which accounted for 70% of the total variance (See Table 5.7). The scree plot also suggests 4 factors, as shown in Figure 5.2. The communalities and factor loadings for retained items are presented in Table 5.8 and Table 5.9 respectively.

Table 5.6 KMO and Bartlett's Test for Final EFA

Kaiser – Meyer – Olkin Measure of Sampling Adequacy.		.902
Bartlett's Test of Sphericity	Approx. Chi – Square	2166.22 **
	df	105

Note. ** : $p < .01$.

Table 5.7 Total Variance Explained by Four Factors in Final EFA

Factor	Initial Eigenvalues			Extraction Sums of Squared Loadings		
	Total	% of Variance	Cumulative %	Total	% of Variance	Cumulative %
1	6.757	45.047	45.047	6.369	42.459	42.459
2	1.575	10.497	55.544	1.169	7.792	50.251
3	1.218	8.118	63.662	.807	5.378	55.629
4	1.050	6.998	70.660	.657	4.377	60.007

Table 5.8 Communalities for Final EFA

	Initial	Extraction
BGS2 I treated ** with a bad attitude	.585	.627
BGS3 I was not considerate enough to **	.616	.638
BGS4 I think I should not be happy anymore since ** has passed away	.401	.505
BGS6 I feel guilty for my own survival following his/her death	.488	.581
BGS7 I did not try everything I could to improve our relationship	.399	.427
BGS8 There are many things I did not do for **	.525	.615
BGS10 I did something hurtful to him/her	.589	.575
BGS13 I feel ** gave me a lot of good things, yet I could not reciprocate	.462	.557
BGS14 I made ** unhappy	.595	.616
BGS16 I feel responsible for his/her death	.660	.717
BGS17 He/she suffered because of me	.663	.683
BGS18 I feel sorry toward ** whenever I am happy after he/she died	.510	.637
BGS22 He/She might still be alive if I had done a better job	.579	.660
BGS25 I think he/she would not have died at that time if I had done things differently	.583	.646
BGS29 I did not spend enough time with **	.430	.516

Note. ** represents the deceased family member.

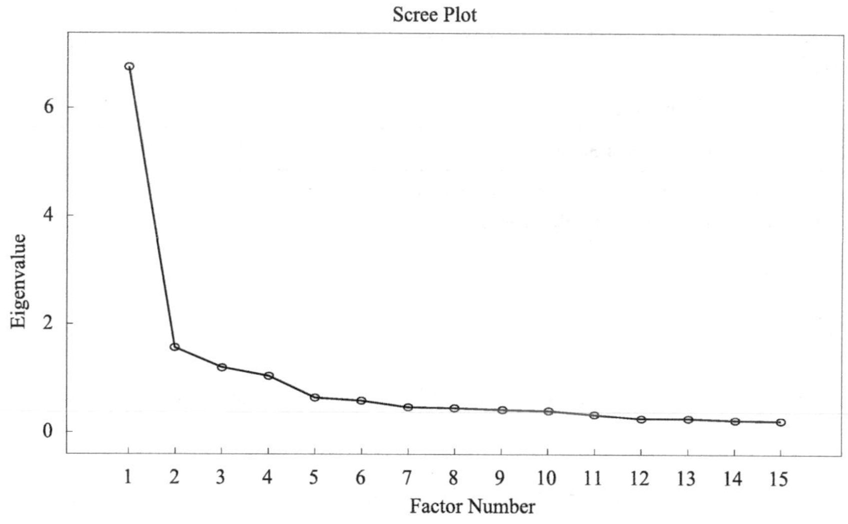

Figure 5.2 Scree Plot for Final EFA

Table 5.9 Rotated Factor Loadings in Final EFA

	Factor			
	1	2	3	4
BGS25 I think he/she would not have died at that time if I had done things differently	.791			
BGS22 He/She might still be alive if I had done a better job	.790			
BGS16 I feel responsible for his/her death	.767			
BGS17 He/she suffered because of me	.615			
BGS2 I treated ** with a bad attitude		-.813		
BGS3 I was not considerate enough to **		-.755		
BGS14 I made ** unhappy		-.709		
BGS10 I did something hurtful to him/her		-.658		
BGS7 I did not try everything I could to improve our relationship		-.555		
BGS18 I feel sorry toward ** whenever I am happy after he/she died			.753	
BGS4 I think I should not be happy anymore since ** has passed away			.709	
BGS6 I feel guilty for my own survival following his/her death			.691	
BGS13 I feel ** gave me a lot of good things, yet I could not reciprocate				.664
BGS29 I did not spend enough time with **				.598
BGS8 There are many things I did not do for **				.597

Note. Factor loadings less than 0.4 are not displayed in the table.

** represents the deceased family member.

Further item analysis

In order to ensure item homogeneity (ie. items in the same scale measures something in common), the item – total correlation was analyzed. The item will be removed if its correlation to the total scale score is less than. 3. All corrected item – total correlations were higher than. 4 (see Table 5. 10), so no item was taken out in this process.

Table 5.10 Item – Total Correlation

	Corrected Item – Total Correlation
BGS8 There are many things I did not do for **	.654
BGS13 I feel ** gave me a lot of good things, yet I could not reciprocate	.548
BGS29 I did not spend enough time with **	.563
BGS2 I treated ** with a bad attitude	.546
BGS3 I was not considerate enough to **	.625
BGS7 I did not try everything I could to improve our relationship	.531
BGS10 I did something hurtful to him/her	.634
BGS14 I made ** unhappy	.627
BGS4 I think I should not be happy anymore since ** has passed away	.493
BGS6 I feel guilty for my own survival following his/her death	.579
BGS18 I feel sorry toward ** whenever I am happy after he/she died	.579
BGS16 I feel responsible for his/her death	.709
BGS17 He/she suffered because of me	.753
BGS22 He/She might still be alive if I had done a better job	.629
BGS25 I think he/she would not have died at that time if I had done things differently	.648

Note. ** represents the deceased family member.

On one hand item homogeneity is desired, and on the other hand item redundancy should be avoided, in order to maximize the breadth of measurement, and to keep an optimal scale length. The selected item should load high on the factor representing the scale, while at the same time exhibit moderate to low item inter – correlations (Boyle, 1991). The optimal item inter – correlations are expected to be lower than. 7 to avoid item redundancy (Corley, Elswick, Gorman, & Clor, 2000). The result found 3 pairs of items (BGS2 and BGS3; BGS10 and BGS 14; BGS 17 and BGS 16) with correlations higher than 0.7. In each pair, the item having a higher factor loading factor was retained and the one with a lower loading was removed. This process results in the removal of BGS3, BGS10 and BGS17.

Table 5.11 Inter – Item Correlations

	BGS8	BGS29	BGS2	BGS3	BGS7	BGS10	BGS14	BGS4	BGS6	BGS18	BGS16	BGS17	BGS22	BGS25
BGS13	.593	.525	.215	.312	.167	.333	.377	.303	.378	.398	.361	.432	.415	.333
BGS8		.566	.299	.390	.326	.433	.415	.335	.454	.388	.438	.504	.478	.466
BGS29			.237	.314	.251	.367	.380	.267	.354	.329	.354	.449	.468	.416
BGS2				.730	.496	.480	.535	.240	.247	.267	.412	.442	.277	.327
BGS3					.493	.517	.557	.299	.302	.277	.440	.514	.372	.402
BGS7						.509	.436	.304	.360	.330	.388	.401	.295	.308
BGS10							.718	.293	.364	.313	.443	.497	.346	.373
BGS14								.271	.288	.294	.436	.494	.366	.345
BGS6									.528	.573	.351	.381	.246	.305
BGS18										.597	.425	.428	.309	.414
BGS16											.479	.442	.375	.379
BGS17												.754	.619	.640
BGS22													.591	.616
BGS25														.684

Note. All correlations are significant($p < .01$).

Summary of the factors related to guilt cognition

In the end, there were 12 items loading on 4 factors, which represent different aspects of guilt cognitions. Each factor contains 3 items, and each item loading was higher than 0.5. Factor 1 reflected guilt about not being able to prevent the death, or attribution of the death to oneself (e. g. "I think he/she would not have died at that time if I had done things differently"). This factor was labeled "responsibility for death". Factor 2 was associated with wrong doings which had hurt the deceased within their past relationship (e. g. "I made ** unhappy"). This factor was labeled "hurting the deceased". Factor 3 contained items expressing guilt about continuing to live or enjoying one's life after the death (e. g. "I feel sorry toward ** whenever I am happy after he/she died"), and was labeled "survivor guilt". The last factor was constructed using items reflecting guilt about the failure to reciprocate toward the parents or failing to do the things one feels obligated to do for the deceased. (eg. "I feel ** gave me a lot of good things, yet I could not reciprocate"), and was thus labeled "Indebtedness guilt".

Analyzing items in guilt feelings

The 3 items in guilt feelings were analyzed using inter – item correlation and item – total correlation to check for internal consistency. The results indicated that one reverse scored item (BGS5) had very low correlations with the other 2 items (.136 and .070 respectively), and was therefore removed from the scale. Since there are only 2 items in this factor, even though the correlation between these items are higher than 0.7, none of them were removed from the scale. This resulted in 2 items being retained in the final guilt feelings sub – scale. The Inter – Item correlations and Item – Total correlations are presented in Table 5.12 and Table 5.13.

Table 5.12 Inter – Item Correlation of Guilt Feeling Items

	BGS5 My guilt does not induce much distress on me (Reverse scored)	BGS21 My heart hurts when recalling things for which I feel guilty toward **	BGS19 I feel upset when I think about things I should have done differently
BGS5	1.000	.136	.070
BGS21		1.000	.744

Notes. ** represents the deceased family member.

Table 5.13 Item – Total Correlation of Guilt Feeling Items

	Corrected Item – Total Correlation
BGS5 My guilt does not induce much distress on me (Reverse scored)	.110
BGS21 My heart hurts when recalling things for which I feel guilt toward **	.664
BGS19 I feel upset when I think about things I should have done differently	.611

Notes. ** represents the deceased family member.

Confirmatory Factor Analysis(CFA)

Confirmatory Factor Analysis is used to test how the proposed structures and organizations of the scale are supported by actual data. A series of CFA were conducted using AMOS 16.0 to compare the one – factor model with multidimensional models of bereavement guilt. The hypothesized models are specified below.

Competing models

Theoretical Model 1(Figure 5.3) : The experience of guilt was not organized into distinct components, and guilt is shown as a one dimension construct. In this model all 14 items were retained after the EFA and other item analysis procedures were loaded on one factor; namely, bereavement guilt.

Theoretical Model 2(Figure 5.4) : In this model, bereavement guilt comprises five components, which reflect different aspects of the concept. Those factors are in the same level, even though one of them describes guilt feelings and the other four factors are about guilt cognition. In this model, items were expected to load on 5 factors; specifically, responsibility for death, hurting the deceased, survivor guilt, indebtedness guilt and guilt feelings.

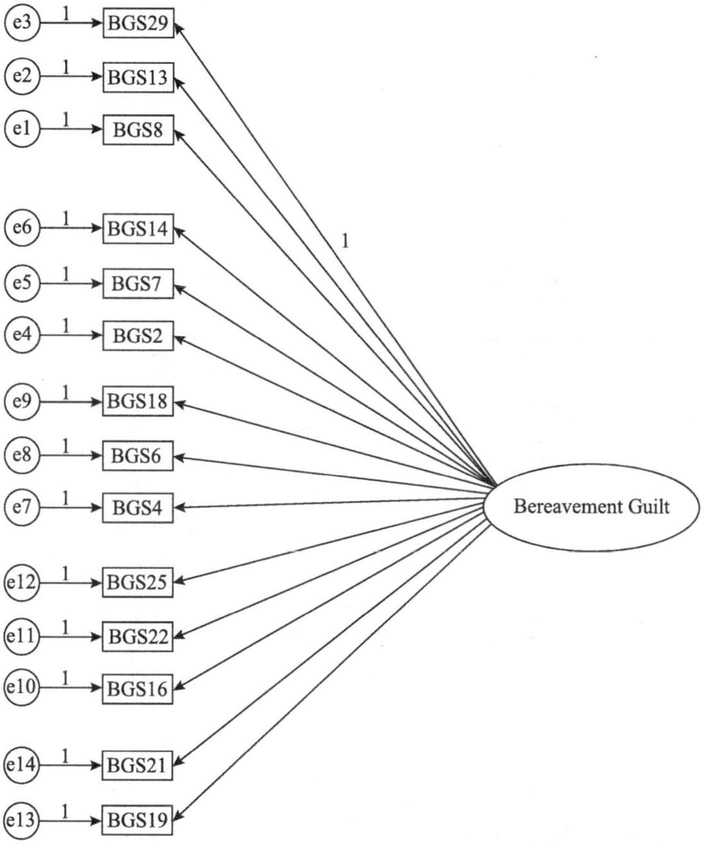

Figure 5.3 Theoretical Structure of Bereavement Guilt Model 1.

Original item number was displayed in the model.

Theoretical Model 3(Figure 5.5):In this model, the bereavement guilt is a second order construct. It comprises both guilt cognition and guilt feelings, while differing content in guilt cognition is reflected by 4 factors. In this model, responsibility for death, hurting the deceased, survivor guilt and indebtedness guilt were loaded on a higher order latent factor labeled guilt cognition. The other two items loaded on guilt feelings. It is supposed that guilt cognition and guilt feelings correlate with each other.

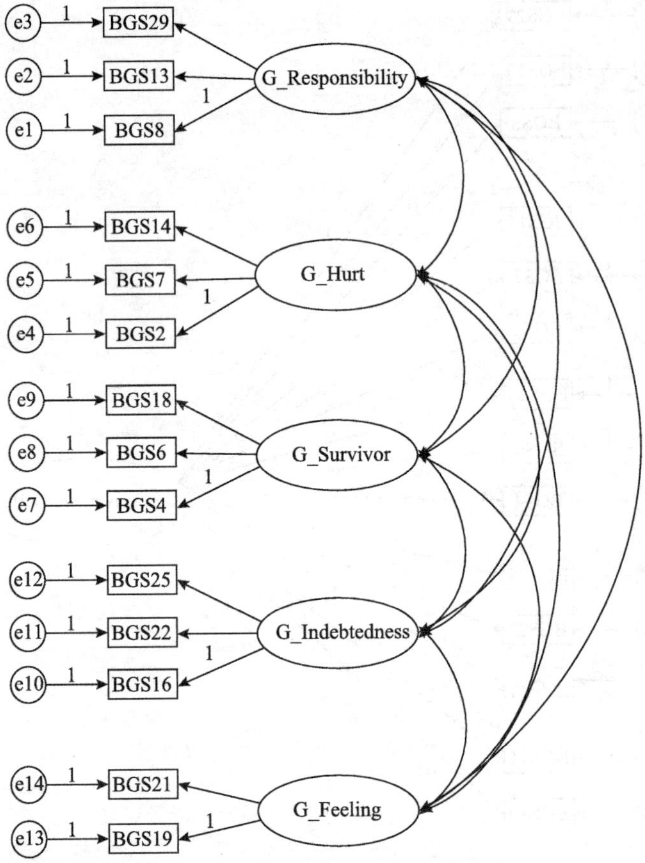

Figure 5.4 Theoretical Structure of Bereavement Guilt Model 2.

G_Responsibility = responsibility for the death; G_Hurt = Hurting the deceased; G_Survivor = Survivor guilt; G_Indebtedness = Indebtedness guilt; G_Feeling = Guilt feelings.

Sample size

There is not an absolute rule regarding the sample size for CFA, and several considerations are suggested by researchers when it comes to the sample issue; such as the estimation technique, model complexity, missing data and communality of items. For example, a widely accepted ratio to minimize the problems with

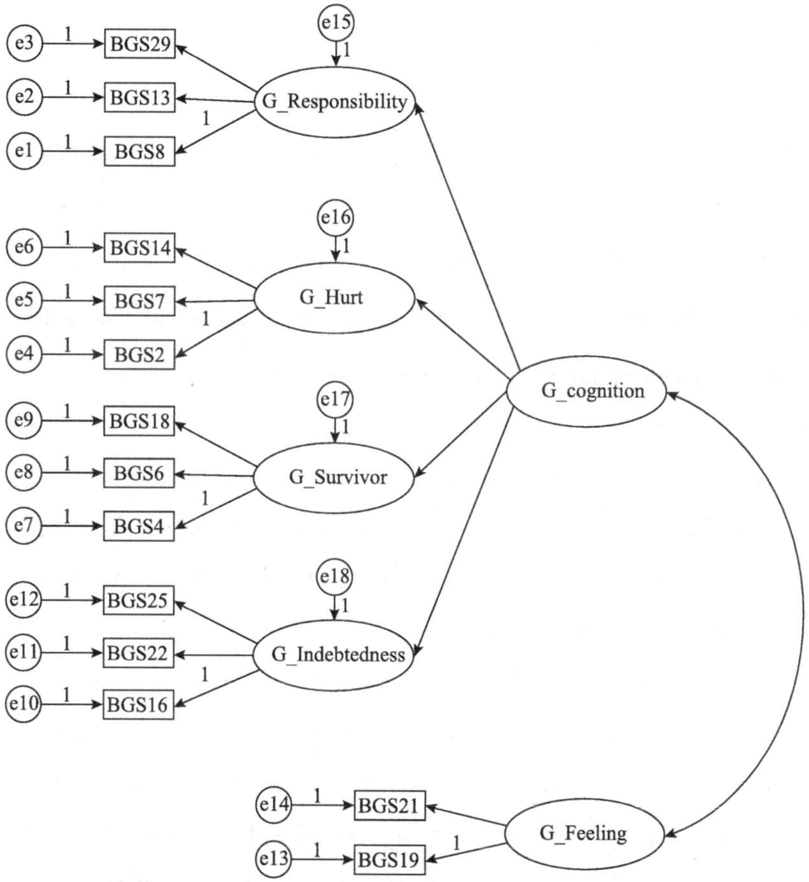

Figure 5.5 Theoretical Structure of Bereavement Guilt Model 3.

G_Responsibility = responsibility for the death; G_Hurt = Hurting the deceased; G_Survivor = Survivor guilt; G_Indebtedness = Indebtedness guilt; G_Feeling = Guilt feelings; G_Cognition = Guilt cognition.

non – normal data is 15 respondents for each parameter estimated in the mode (Wang, Fan, & Wilson, 1996). Based on this calculation (Model 3 with the most parameters needing to be estimated, 33 × 15 = 495), around 500 cases are needed. The present study uses the remaining data from 600 cases, which is sufficient to satisfy the minimum sample size requirement.

The model fit indices

Multiple goodness - of - fit indices should be included to evaluate the model, as different indices provide different kinds of fit information. Generally, there are 3 kinds of fit indices. Absolute fit indices are direct measures to indicate how well the proposed model fits the observed data. The most basic absolute fit index is χ^2. The significant χ^2 test means that there are significant differences between the hypothesized model and the observed data. However, χ^2 is highly influenced by sample size, and the test usually yields statistically significant results. Therefore, it is not the most informative index, and other absolute fit indices are usually reported to provide further information of the model fit. The most commonly used indices include the Goodness - Of - Fit Index (GFI), Root Mean Square Error of Approximation (RMSEA), and Normed Chi - Square (the ratio of Chi - Square to the degrees of freedom). Another type of index is incremental fit indices, which compare the proposed model with alternative baseline models, in terms of how well they fit the observed data. The most commonly used base line model is the null model, which assumes no correlation between any observed variables. Therefore, the incremental fit indices indicate improvement through specifying loadings between item and construct, and through correlations between variables. The most widely used indices in this type include the Normed Fit Index (NFI), the Tucker Lewis Index (TLI), and the Comparative Fit Index (CFI). The third cluster of indices, called Parsimony Fit Indices, takes into account the model complexity when comparing competing models. For example, the Adjusted Goodness of Fit Index (AGFI) is the adjustment value of the GFI based on the ratio of degrees of freedom in one model to total degrees of freedom. A better parsimony fit index suggests either a better fit or a simpler model.

There are different opinions related to the cut - off value of model fit indices to indicate an acceptable model (Bentler, 1990; Hu & Bentler, 1999; Marsh, Balla, & McDonald, 1988). However, researchers seem to agree that multiple indices of different types should be used, rather than single index, and the value of acceptable model fit indices should take into account model characteristics; such as sample size and observed variables. Following up these guidelines, this research reported multiple fit indices and has compared 3 proposed models. The following criteria were adopted to indicate acceptable indices: $\chi^2/df < 3$, lower bound of the 90% Confident Interval for the RMSEA <.05, and upper bound <.10, CFI >.90 (Kline, 2005), TLI and NFI >.90 (Bentler, 1990), AGFI >.8 (Marsh et al., 1988). While it may be difficult to assert the absolute fit of the model, it is easier to compare models and identify the best one. When comparing models, the improvement of a nested model will be assessed by the χ^2 difference test.

Table 5.14 Model Fit Indices for Three Competing Models

	χ^2	df	χ^2/df	GFI	AGFI	NFI	TLI	CFI	RMSEA(90% CI)
Model 1	1073.457 **	77	13.941	.766	.680	.754	.724	.767	.147(.139 - .155)
Model 2	210.60 **	67	3.143	.951	.923	.952	.954	.966	.060(.051 - .069)
Model 3	249.37 **	72	3.463	.941	.914	.943	.947	.958	.064(.056 - .073)

Note. ** : $p < .01$.

The goodness of fit coefficients for the three models is presented in Table 5.14. Model 1 did not fit the data well, as all the indices fail to reach acceptable values. Both model 2 and model 3 demonstrated a good fit with the data. The consequent χ^2 difference test suggested significant improvement in model 2 compared with model 3 ($\chi^2 = 38.77$, $\Delta df = 5$, $p < .01$). This suggests that even though both model 2 and model 3 are acceptable, model 2 is superior to model 3. This also supports the idea that bereavement guilt contains both cognitive and affective components, but compared to the hierarchical organization, the first-order structure of guilt in bereavement may be closer to the actual representation of this experience in bereaved people. Based on model 2, the correlations among factors are presented in Table 5.15.

Table 5.15 Correlations Among Factors in Model 2

	Responsibility for death	Hurting the deceased	Survivor guilt	Indebtedness guilt
Hurting the deceased	.53			
Survivor guilt	.58	.56		
Indebtedness guilt	.63	.63	.66	
Guilt feeling	.80	.57	.73	.68

Construct validity

The correlation between the test scores of the developed scale with scores from other tests, in accordance with the theory predictions, will provide evidence of the construct validity of the new scale. The whole sample was intended for use in assessing the construct validity for the scale. However, 51 of the informants failed to complete the self – esteem scale, so therefore only 822 cases were included in the following statistics.

Correlation with Self – esteem and self – forgiveness

The negative correlation between guilt and self – forgiveness, as well as self – esteem, provided supporting evidence of the construct validity of the Bereavement Guilt Scale (Table 5.16).

Table 5.16 Correlations of the BGS with Self – Esteem and Self – Forgiveness

	Responsibility for death	Hurting the deceased	Survivor guilt	Indebtedness guilt	Guilt feeling	Guilt in total
Self – forgiveness	–.364 **	–.287 **	–.363 **	–.376 **	–.440 **	–.458 **
Self – esteem	–.290 **	–.385 **	–.339 **	–.206 **	–.296 **	–.375 **

Note. ** : $p < .01$.

Correlation with depression and anxiety

The association between guilt measured by the Bereavement Guilt Scale and depression and anxiety scores was examined. The results support the belief that the Bereavement Guilt Scale has good validity, given the significant positive correlation between each sub – scale, as well as between the total score of the whole scale and depression/anxiety (Table 5.17).

Table 5.17 Correlations of BGS with Depression and Anxiety

	Depression	Anxiety
Responsibility for death	.279 **	.263 **
Hurting the deceased	.311 **	.263 **
Survivor guilt	.505 **	.391 **
Indebtedness guilt	.253 **	.217 **
Guilt feeling	.352 **	.302 **
BGS total	.421 **	.356 **

Note. ** : $p < .01$.

Reliability

The reliability of the scale was assessed by internal consistency reliability and test – retest reliability (see Table 5.18). The Bereavement Guilt Scale (BGS) is found to have good internal consistency, as all of the Cronbach's α of each sub – scale and of the total scale was above .7. Taking into consideration that there are only 2 to 3 items in each sub – scale, this result is sufficient to ensure its internal consistency. Moreover, 24 participants filled out the BGS again 3 to 5 weeks after the first time they filled out the questionnaire. The correlation between their first scores and second scores ranged between .722 and .931. The result demonstrated the stability of BGS scores across time, suggesting good test – retest reliability.

Table 5.18 Internal consistency and test – retest reliability of BGS

BGS	Cronbach's α		Test – retest correlation
	Sample 2	Sample 3	
Responsibility for death	.85	.88	.84
Hurting the deceased	.74	.74	.90
Survivor guilt	.80	.82	.73
Indebtedness guilt	.79	.75	.83
Guilt feeling	.85	.86	.72
BGS total	.88	.92	.93

Discussion

The content of the BGS

The development process of the BGS integrated both deductive and inductive paradigms. The conceptualization of bereavement guilt guides the creation of the item pool to include both cognitive and affective dimensions. However, the conceptualization of guilt did not specify particular content or factors in guilt cognition, suggesting only that there are multiple sources in this aspect of guilt. The factors in guilt cognition and the structure of bereavement guilt were generated from the empirical data collected in this study. Items in this scale were all translated from bereaved people's narratives about their own experiences of guilt in bereavement. Therefore they truly reflect the essence of this phenomenon, and ensure the validity of the scale. The BGS contains 5 sub – scales, reflecting bereaved adult children's guilt about being responsible for the death, hurting the deceased, surviving after the death, and indebtedness to the deceased, as well as the distressful feeling accompanied by those guilt thoughts. This scale comprises types of guilt that were mentioned widely in previous literature, such as responsibility for the death and survivor guilt. At the same time it also includes aspects that were less noticed or assessed in previous studies, such as hurting the deceased and indebtedness guilt. Moreover, it includes one factor that depicts the degree of distress brought about by the guilt feelings, which was neglected in most previous investigations. Both the interview data and the survey data suggest that this content in guilt should not be ignored.

Psychometric property of the BGS

The dimensionality and construct validity of the BGS have been evaluated with a series of statistics, and all results suggest the good validity of this scale. Its dimensionality has been supported by both EFA and CFA, and the direction and magnitude of the correlation between BGS scores and relevant psychological constructs, including self – forgiveness, self – esteem, depression and anxiety, are in accord with theoretical predictions. The BGS has also demonstrated good internal consistency and high test – retest reliability. Taken together, it makes a valid and reliable instrument by which to assess bereavement guilt. Researchers can use its total score to indicate general guilt, while at the same time assessing the situation of different guilt content through sub – scale scores, given the homogeneity as well as dimensionality of the BGS, as supported by the data.

The structure of bereavement guilt

The findings of the CFA confirmed the multidimensional conceptualization of bereavement guilt, because the multiple factor models fit the observed data much better than did the single factor model. Results also suggested that the one factor model of bereavement guilt did not fit the data. Both the first – order five factor model and the 5 factors loading on two second orders model revealed acceptable model fit indices; however, the follow up analysis indicated that the first – order model is superior to the second – order model. Statistically, this may be due to the fact that model 2 is simpler than model 3, and thus fits better with the parsimonious principle of model testing, yielding better model fit coefficients. Theoretically, on one hand, this finding supports the belief that bereavement guilt contains two main aspects; namely, guilt cognition and guilt feelings. On the other hand, it also suggests that the representation of the guilt experience among bereaved people may not be separated in such a neat and tidy manner as guilt cognition and guilt feelings, following subsequent differentiation within guilt cognition. Instead, in their bereavement guilt experiences, all kinds of guilt cognitions and the accompanying distressful feelings exist all at the same time.

Limitations

The development and validation process followed suggestions from experts on scale development, using EFA and CFA in the scale validation process. For example, a large sample size for EFA and CFA analysis increases the chance to generate reliable statistic results. However, limitations still exist. The participants in this study were recruited from memorial websites; therefore, it is not a completely random sampling approach. There is a lack of evidence to support differences in characteristics between memorial web – sites users and those who seldom use such sites. However, one study collected data using both on – line surveys and a traditional paper and pencil method, and compared the nature of those two samples (Tolstikova, 2009—2010). This found no difference in the degree of grief and in most demographic variables, except that the on – line survey recruited a more demographically diverse sample. In the recent study, the analysis on participants' background (demographic and death related) factors demonstrated the diversity of the participants; in terms of their age, gender, education, marital status, religion, bereavement time, and information of the deceased. However, participating in the research was a self – selection process. The introduction of the study and informed consent form states that this study aimed to understand individuals' grief experiences. Whether or not this survey attracted bereaved people who were more reflective, or had experienced more intensive grief than those who did not participate, it is impossible to tell at this time. The possibility exists, but no empirical evidence has been gained so far to answer this question.

Summary

Based on previous conceptualization of bereavement guilt and qualitative data collected from the first study, a scale (BGS) which was specifically designed to assess bereavement guilt in adult children was developed and validated. This scale found that:

The BGS contains 14 items, loading on five factors, namely, responsibility for death, hurting the deceased, survivor guilt, and indebtedness guilt.

The BGS is proven to have good face validity, factorability and constructive validity.

The BGS is proven to have good internal consistency and test – retest reliability.

Bereavement guilt is a multidimensional construct, rather than a one – dimensional concept.

There are both cognitive and affective aspects in bereavement guilt, and four types of guilt cognitions are identified.

Both the one order multidimensional model and the second order two aspects model were supported by empirical data, but the one order structure of bereavement guilt is superior to the hierarchical structure.

In summary, this is the first validated scale to measure guilt in the context of bereavement. This instrument bridges the theories on guilt with empirical investigations. As will be illustrated in the next chapter, the BGS facilitates further understanding of bereavement guilt at the population level, including its prevalence, predictors and relationship to grief symptoms.

Chapter 6

Bereavement guilt: Its prevalence, predictors and relationship to complicated grief

Introduction

As reviewed in Chapter 2, previous findings related to guilt in bereavement have involved three main themes: prevalence, risk factors and its relationship with health outcomes.

Guilt is reported to be one of the main themes in people's grief experience. Researchers and clinicians have suggested that guilt is a common emotion experienced by bereaved people. The question of how common guilt is among bereaved people should be answered by quantitative studies which can investigate the percentage of people who have experienced guilt under these circumstances. However, the answers provided by previous quantitative studies show much discrepancy in their results, and guilt prevalence varies over enormous ranges. One underlining reason is the lack of a standardized measurement of guilt, and that researchers have used different indicators to assess guilt. Therefore, different studies reported the prevalence of different types of guilt, in various subgroups of people. The Bereavement Guilt Scale now provides a new solution to this problem, as it can be used to explore not only the prevalence of general guilt but also different components of bereavement guilt.

Another theme seen in previous studies involving bereavement guilt is to examine the risk factors for higher guilt levels. In general, sparse efforts have been made in this area. A relatively greater number of studies have discussed the role of gender, bereavement time and cause of the death than have discussed other risk factors. In spite of this, inconsistent findings have also been documented in previous studies on these three factors. Exploring risk factors can not only contribute to the knowledge and theory related to this phenomenon (W. Stroebe & Schut,

2001), but also has important clinical implications. The findings regarding risk factors can help identify susceptible people, so that clinicians can be aware of their vulnerability and deliver early support and intervention (M. S. Stroebe, Folkman, Hansson, & Schut, 2006). In this study, the demographics as well as some contextual variables will be investigated based on their associations with bereavement guilt. Since those risk factors are unchangeable factual variables, I use the terms *risk factors* and *predictors* interchangeably to refer to them in this research.

Most previous bereavement studies that explore risk factors have the shortcoming of only assessing one factor at a time. The simple correlation between risk factors and outcome variables can be confusing because this does not eliminate other confounding variables. For example, one study which found that losing a boy induces higher guilt on bereaved parents than losing a girl, may be due to the fact that more boys in the study sample died from unnatural deaths than natural deaths (Hazzard et al., 1992). Therefore it is possible that the gender difference of the deceased child actually reflects the impact of cause of death on the level of guilt. This shortcoming of first order correlation analysis in investigating risk factors has been noticed by some researchers (M. S. Stroebe et al., 2006; van der Houwen et al., 2010; Wijngaards - de Meij et al., 2005), and a multivariate analysis approach is preferred. Therefore, in the present study, a multivariate approach will be used, in addition to simple regression, which concludes single predictors in the analysis.

Another important theme that has attracted researchers' attention is the health outcomes of bereavement guilt. The review in chapter 2 shows that there have been considerable studies indicating that guilt is associated with negative health outcomes, both psychologically and physically. However, only a few of those studies assess the relationship between guilt and grief reactions, whereas most of them focus on other psychological distresses, such as depression and anxiety.

Moreover, only 3 studies have explored the relationship between guilt and complicated grief, and 2 of them found a positive association (Boelen & Lensvelt - Mulders, 2005; Boelen et al., 2003) while the other one did not find a significant contribution of guilt to complicated grief (Golden & Dalgleish, 2012). Complicated grief has been proved to be distinct disorder from depression and anxiety (Boelen & van den Bout, 2005; Prigerson & Frank, 1995). Exploring the relationship between guilt and complicated grief will not only enrich the theoretical understanding of factors that are associated with pathological grief reactions, but also provide insight in treating complicated grief. If the positive relationship between bereavement guilt and complicated grief is established, the important role of guilt in adjusting to grief will be further established.

Research objectives:

With the newly developed instrument, this study aims to

1) discover the prevalence of bereavement guilt,

2) investigate demographic and death related predictors of guilt, and

3) explore the association between bereavement guilt and complicated grief.

Specifically, the research questions in this chapter are:

How frequently is this kind of guilt felt by bereaved Chinese adult children in general?

What demographic and death related factors predict higher bereavement guilt in total?

What demographic and death related factors predict higher levels of each type of bereavement guilt?

What is the relationship between bereavement guilt and complicated grief?

What is the relationship between each type of bereavement guilt and complicated grief?

Given the problematic measurement of guilt, and the limited and discrepant findings from previous studies, this initiative is exploratory in nature.

Method

Sample and questionnaires

The same data set comprising same 822 bereaved adult children introduced in Chapter 5 was analyzed in this study. Besides the questionnaires listed in chapter 5, the Inventory of Complicated Grief was also included in the scale package. The Inventory of Complicated Grief (ICG) contains 19 items, with a self – reported statement about traumatic distress and separation distress. The test – retest reliability is .80 over 6 months of loss, and the internal consistency was originally reported as .94. The scale displays good concurrent validity and discriminating validity (Prigerson et al., 1995). The Cronbach alpha of ICG in the present sample is .932.

The data collection methods and participant information has been stated in Chapter 5. The data was analyzed in this study using PASW Statistic 18.

Data analysis plan

Prevalence of bereavement guilt

As the BGS is a five point scale (1 = does not describe me very well, 2 = does not quite describe me, 3 = describes me fairly well, 4 = describes me well, 5 = describes me very well), experiencing guilt is operationalized as obtaining a mean score on the sub – scale or on the total scale, equal to or above 3. The prevalence of guilt equals the number of individuals who experienced guilt divided by the total number of participants.

The prevalence in the unnatural death group will be compared with the natural death group.

Predictors of bereavement guilt

The predicting effect of each risk factor will be analyzed with both simple regression, for first order correlation, and multiple regression, for controlling shared variance among risk factors.

The demographic risk factors investigated in this study include: age, gender, education level, marital status and religious belief. Risk factors related to information of the deceased include age and gender of the deceased parents. Death related risk factors include time passed since the death, cause of the death, and the bereaved person's psychological preparation for the death.

Relationship between bereavement guilt and complicated grief

Both the relationship between total guilt and complicated grief and between each guilt component and complicated grief, will be analyzed with correlation analysis and regression analysis. The mediating role of guilt feelings between guilt cognition and complicated grief will be assessed by a series of regression analysis.

Results

Prevalence

In order to investigate guilt prevalence, the mean score of each BGS sub – scale and of the whole BGS scale was calculated for every participant. Accordingly, the prevalence of each aspect of guilt and of total guilt was calculated and presented in Table 6.1. The prevalence varies across different types of guilt, ranging from 23% to 81%. Indebtedness guilt and guilt feelings are the two most widely reported components of guilt.

Table 6.1 Prevalence of Guilt in Total Sample

	Responsibility for death	Hurting the deceased	Survivor guilt	Indebtedness guilt	Guilt feeling	Guilt in total
N(Mean > =3)	306	187	219	668	576	325
Percentage	37%	23%	27%	81%	70%	40%

Note. Total number of participants is 822.

The mode of death is expected to influence the guilt experienced by bereaved people (Rando, 1993). There are studies suggesting that guilt is more common in people who are bereaved due to unnatural death than those who are bereaved from a natural death (Jordan, 2001). Therefore, different guilt prevalence is expected among people bereaved due to different causes, and the prevalence of guilt was calculated for these two sub – groups. As shown in Table 6.2, the prevalence of guilt (regardless of guilt type or total guilt) was higher in the unnatural death group (including suicide, accident, unknown reason and other non illness causes of death) than in the natural death group (including chronic illness and sudden illness). Again, the most commonly experienced guilt is indebtedness guilt, which was experienced by as many as 96.8% of adult children who had lost their parents from unnatural causes, and 80.9% of the same group of people reported that their guilt was painful and distressful. The Chi – square test indicated that the differences between the natural and unnatural death groups were significant on all types of guilt except survivor guilt.

Table 6.2 Prevalence of Guilt in Different Subgroups

	Responsibility for death	Hurting the deceased	Survivor guilt	Indebtedness guilt	Guilt feelings	Guilt in total
Natural death (N = 728)	35.9%	22.2%	22.4%	79.3%	68.8%	37.7%
Un – natural death (N = 94)	47.9%	30.9%	33%	96.8%	80.9%	54.3%
χ^2	5.15 *	4.10 *	2.18	16.84 **	5.88 *	9.72 **

Note. * : $p < .05$; ** ; $p < .01$.

Demographic and contextual risk factors

Simple regression analysis

For each component of guilt and for total guilt, a simple regression was conducted to investigate the predicting effects of each risk factor. Each simple regression model includes one predictor. The continuous independent variables were directly included in the regression model, while the categorical variables were coded into dummy variables before being entered into the regression model. The results are summarized in Table 6.3.

Regarding guilt about responsibility for the death, younger ages of the bereaved adult children, younger ages of the deceased parents and less psychological preparation all predict a higher score on this sub – scale. Meanwhile, the following factors also predict a higher risk of feeling responsible for the death: losing a mother, compared to losing a father; bereavement from an unnatural death rather than a natural death; being single rather than married; and having a religion, compared to having no religious belief.

Regarding guilt over hurting the deceased, most of the risk factors were the same as guilt about responsibility for the death, except that the gender of the deceased parents as well as religious beliefs ceased to play a significant role in predicting the levels of this kind of guilt experienced by people.

Survivor guilt shared the most risk factors with guilt about responsibility for the death; however, the cause of the death and marital status of the bereaved child does not influence the intensity of this type of guilt.

The risk factors that predict higher responsibility for the death also predict higher levels of indebtedness guilt; with the exception of religion, which did not show significant influence on indebtedness guilt. Meanwhile, the gender of the bereaved person did influence this type of guilt: women scored higher than men on this dimension.

As for guilt feelings, most risk factors that predicted indebtedness guilt also predict this dimension, except for marital status, which did not predict guilt feelings. It's worthy to note that, guilt feelings is the only aspect of bereavement guilt that showed a mild trend of decreasing over time. Even though the regression coefficient was small(-.082), it was statistically significant.

Finally, the age of the bereaved child, the age and gender of the deceased parents, psychological preparation for the death, the cause of the death, and the marital status of the bereaved all significantly predict higher total scores of guilt in the simple regression analysis.

Table 6.3 Standardized Coefficients of Predictors of Guilt in Simple Regression Analysis

	Responsibility for death	Hurting the deceased	Survivor guilt	Indebtedness guilt	Guilt feeling	Guilt in total
Age of the bereaved	-.275 **	-.281 **	-.155 **	-.322 **	-.208 **	-.313 **
Age of the parents	-.199 **	-.233 **	-.119 **	-.307 **	-.173 **	-.259 **
Gender of the bereaved (Men = 0)	.010	-.020	.042	.116 **	.117 **	.062

to be continued

	Responsibility for death	Hurting the deceased	Survivor guilt	Indebtedness guilt	Guilt feeling	Guilt in total
Gender of the parents (Father = 0)	.110 **	−.072	.118 **	.113 **	.086 *	.126 **
Bereavement Time	−.066	.006	−.056	−.038	−.082 *	−.058
Psychological preparation	−.152 **	−.091 **	−.155 **	−.187 **	−.136 **	−.182 **
Cause of the death (Natural death = 0)	.093 **	.094 **	.067	.193 **	.098 **	.136 **
Marital status (single = 0)						
Married	−.074 *	−.149 **	−.017	−.087 *	−.060	−.096 **
Lose Spouse	−.023	.003	.0003	−.033	−.025	−.019
Religion (No religion = 0)	.090 *	.046	.001	.043	.021	.053
Education	.052	.020	−.027	.044	.023	.029

Note. *: $p < .05$; **: $p < .01$.

Table 6.4 Standardized Coefficients of Predictors of Guilt in Multivariate Regression Analysis

	Responsibility for death	Hurting the deceased	Survivor guilt	Indebtedness guilt	Guilt feeling	Guilt in total
Age of the bereaved	−.265 **	−.226 **	−.172 **	−.205 **	−.146 **	−.259 **
Age of the parents	.012	−.030	.010	−.110 *	−.040	−.038
Gender of the bereaved (Men = 0)	−.027	−.038	.017	.085 *	.096 **	.028
Gender of the parents (Father = 0)	.129 **	.090 **	.134 **	.133 **	.098 **	.148 **
Bereavement Time	−.044	.024	−.040	−.014	−.056	−.032

to be continued

	Responsibility for death	Hurting the deceased	Survivor guilt	Indebtedness guilt	Guilt feeling	Guilt in total
Psychological preparation	-.119 **	-.054	-.138 **	-.108 **	-.090 **	-.129 **
Cause of the death (Natural death = 0)	.022	.029	.007	.099 **	.035	.048
Marital status						
Married	.019	-.081	.070	.007	-.007	.004
Lose Spouse	.018	-.016	.060	-.016	-.016	.009
Religion (No religion = 0) [a]	.073 *	.022	-.007	.032	.015	.036
Education	.022	-.0003	-.057	.013	-.002	-.005
R^2	.112 **	.098 **	.067 **	.167 **	.079 **	.144 **

Note. [a]: As sample sizes of religion groups other than Buddhism were too small, this variable was coded as with or without a religious belief.

*: $p < .05$; **: $p < .01$.

Multivariate analysis

Those risk factors may share some covariance. For example, both younger ages of the deceased parents and younger ages of the bereaved children are found to predict higher levels of guilt. However, it is possible that the age of the children positively correlates with the age of their parents. Whether one of them is a confounding variable, or whether the unique variance of each of them has additional prediction toward guilt when the common variance shared by both of them is controlled can only be answered by the result of a multivariate analysis which includes all these factors into one regression model. Multiple regression analysis was conducted for each dimension of guilt, as well as for total guilt. The results are presented in Table 6.4.

In order to identify the most important core risk factors and eliminate potential confounding variables, a multivariate regression was conducted for each dependent variable.

Three factors; namely, age of the bereaved, gender of the bereaved, and psychological preparation, were found to be significant in the simple regression analysis and continued to show significant influence on each type of guilt (as well as on the total guilt score) in multiple regression. As can be seen from Table 6.4, the younger age of the adult children, losing a mother and less psychological preparation for the death predict higher guilt.

The effect of gender in predicting indebtedness guilt and guilt feelings was also significant, as it was in the simple regression analysis. The same was found for religion, which continued to predict responsibility for death. However, it worth noticed that the regression coefficient of religion to guilt was only less than 0.1 even though statistically the number was significant. Therefore, the predicting power of religion belief on bereavement guilt was actually not strong.

Some factors which had been found to predict all aspects of guilt now changed to predict one type of guilt when other risk factors were taken into consideration together. For example, age of the deceased parents, which was found to predict all types of guilt, as well as total guilt, now only predicts indebtedness guilt. This means that if the parents died at a younger age, the bereaved adult children will feel greater indebtedness guilt toward the parents, but not other types of guilt; regardless of other demographic factors. The same pattern was found on cause of the death. When controlling for other risk factors, unnatural death does not necessarily induce higher guilt. Instead, it only predicts higher indebtedness guilt.

Some predictors changed from significant to insignificant in predicting bereavement guilt, such as bereavement time and marital status.

No variable changed from insignificant in simple regression analysis to significant in multivariate analysis.

Bereavement guilt and complicated grief

Complicated grief was measured by the ICG, which contains one item about survivor guilt ("I feel that it is unfair that I should live when this person has died"). In order to avoid an inflated association between guilt and complicated grief due to similar items in measuring these two constructs (there are items measuring survivor guilt in the BGS), the item of survivor guilt in the ICG was taken out when calculating the sum score of complicated grief.

Correlation between bereavement guilt and complicated grief

The total bereavement guilt score showed moderate correlation with complicated grief: $r = .605, p < .01$.

The incremental predicting effect of bereavement guilt

The incremental predicting effect of bereavement guilt on complicated grief was assessed through controlling background variables and psychological distress; such as depression and anxiety in the hierarchical regression analysis. As shown in Table 6.5 and Table 6.6, the influence of bereavement guilt on complicated grief was significant, going beyond background variables and psychological distress.

Table 6.5 Hierarchical Regression Analysis of Complicated Grief on Background Variables and Bereavement Guilt

Independent variables	Model 1	Model 2
	β	β
Age of the bereaved	$-.179^{**}$	$-.039$
Age of the parents	$-.056$	$-.037$
Gender of the bereaved (Men = 0)	$.087^{**}$	$.072^{*}$
Gender of the parents (Father = 0)	$.148^{**}$	$.068^{*}$

to be continued

Independent variables	Model 1	Model 2
	β	β
Bereavement Time	-.089 *	-.073 *
Psychological preparation	-.227 **	-.157 **
Cause of the death(Natural death =0)	.073 *	.047
Marital status		
Married	.050	.047
Lose Spouse	.086 *	081 *
Religion(No religion =0)	.008	-.016
Education	-.043	-.039
Bereavement guilt		.534 **
R^2	.176 **	.420 **
ΔR^2		.244 **

Note. Dependent variable: Complicated grief.

* : $p < .05$; ** : $p < .01$.

Table 6.6 Hierarchical Regression Analysis of Complicated Grief on Background Variables, Psychological Distress and Bereavement Guilt

Independent variables	Model 1	Model 2	Model 3
	β	β	β
Age of the bereaved	-.179 **	-.100 *	-.010
Age of the parents	-.056	-.066	-.051
Gender of the bereaved(Men =0)	.087 **	.065	.058 *
Gender of the parents(Father =0)	.148 **	.115 **	.065 *
Bereavement Time	-.089 *	-.049	-.052
Psychological preparation	-.227 **	-.144 **	-.117 **
Cause of the death(Natural death =0)	.073 *	.051	.039
Marital status			
Married	.050	.092 *	.076 *
Lose Spouse	.086 *	.031	.044

to be continued

Independent variables	Model 1	Model 2	Model 3
	β	β	β
Religion(No religion = 0)	.008	.014	-.006
Education	-.043	.008	-.005
Depression		.318 **	.188 **
Anxiety		.202 **	.168 **
Bereavement guilt			.416 **
R^2	.176 **	.371 **	.498 **
ΔR^2		.196 **	.127 **

Note. Dependent variable: Complicated grief;
* : $p < .05$; ** : $p < .01$.

The relationship between guilt components and complicated grief

The BGS delineates guilt into different components, making it possible to examine the relationship between each type of guilt with complicated grief, in terms of the intensity of the associations and possible pathways of the influence.

Correlation between guilt components and complicated grief

The correlation analysis showed that each component of guilt is positively correlated with complicated grief (Table 6.7).

Table 6.7 Correlations Between Types of Bereavement Guilt and Complicated Grief

	Responsibility for death	Hurting the deceased	Survivor guilt	Indebtedness guilt	Guilt feeling	Total guilt
Complicated grief	.440 **	.377 **	.662 **	.452 **	.497 **	.605 **

Note. ** : $p < .01$.

Incremental effect of each guilt component on complicate grief

The associations between each guilt component and complicated grief over background variables and psychological distress were assessed by a series of hierarchical regression analysis. In the first level of the regression model, the statistical contribution of background factors, depression and anxiety, were entered; and in the second level, each guilt component was included to investigate the effect of the unique variance of respective guilt components in explaining the variance of complicated grief. As presented in Table 6.8, the association between each guilt component and complicated grief was significant even when the background variables and psychological distress levels were controlled in the regression model.

Table 6.8 Regression Coefficient of Each Guilt Component Controlling for Background and Psychological Distress Variables

	β	R^2	ΔR^2
Independent variables entered in the second level			
Responsibility for death	.358 **	.290 **	.114 **
Hurting the deceased	.311 **	.252 **	.087 **
Survivor guilt	.600 **	.511 **	.336 **
Indebtedness guilt	.351 **	.267 **	.103 **
Guilt feeling	.420 **	.338 **	.163 **

Note. Dependent variable: Complicated grief. The coefficients presented were results in five distinct regression models. Each regression model assesses the contribution of one guilt component after controlling background variables, depression and anxiety.

** : $p < .01$.

Multiple regression analysis on guilt components

In order to further explore the relationship between each component of guilt with complicated grief, all guilt components were entered into the regression on complicated grief. The results, as presented in Table 6.9, showed that survivor guilt and indebtedness guilt had additional contributions to complicated grief beyond other guilt components.

Table 6.9 Multiple Regression Analysis of Bereavement Guilt to Complicated Grief

	B	*SE*	*β*
Independent variables			
Responsibility for death	.286	.145	.073
Hurting the deceased	-.202	.169	-.042
Survivor guilt	2.470	.153	.546 **
Indebtedness guilt	.628	.157	.141 **
Guilt feeling	.324	.234	.055

Note. $R^2 = .468, p < .01$.

** :$p < .01$.

Mediating effect of guilt feelings

The conceptualization of bereavement guilt comprises both cognitive and affective aspects. It is possible that guilt cognitions influence complicated grief through inducing uncomfortable distresses; i. e. , guilt feelings. Guilt cognitions were calculated from summarizing the total scores from the four sub - scales regarding different types of guilt cognition; namely, responsibility for the death, hurting the deceased, survivor guilt and indebtedness guilt. The mediating role of guilt feelings was tested with a serial of regression analysis following standardized steps suggested by Baron and Kenny (1986). This method is one of the most commonly adopted methods to test mediating effects by psychologists (Frazier & Tix, 2004). According to Baron and Kenny's method (1986), all of the following results should emerge to establish the mediating effect. The first is the significant relationship between the predictor and the outcome. The second is significant association between predictor and the mediator. Then, in the final step, both predictor and mediator are included into the regression model. If the regression coefficient of the mediator is significant, and the effect of predictor on dependent variable showed little more than zero, then the completed mediating effect is established. If the relationship between the predictor and the outcome is smaller but still significant, the mediator partially mediates the association between the predictor and the outcome.

As shown in Table 6.10, guilt cognition is significantly associated with complicated grief (in regression model 1) and guilt feeling (in regression model 2). When both guilt cognition and feelings were entered into the regression model, the regression coefficient of guilt feelings was significant, and the effect of guilt cognition reduced (regression model 3). Therefore, guilt feeling partially mediates the relationship between guilt cognition and complicated grief.

Table 6.10 Regression Analysis to Test the Mediating Effect of Guilt Feelings

	Model 1	Model 2	Model 3
	Dependent variables		
	Complicated grief	Guilt feeling	Complicated grief
Independent variables			
Guilt cognition	8.189 **	.745 **	7.085 **
Guilt feelings			1.482 **
R^2	.362 **	.556 **	.367 **
ΔR^2			.005 **

Note. ** : $p < .01$.

The above result is not enough to prove the significance of mediating effect, which depends on whether the regression coefficient of guilt cognition in model 1 significantly differs from its coefficient in model 3 (Frazier & Tix, 2004). This was tested by calculating the Z score of the difference, which equals the product of guilt cognition's regression coefficient in model 2 (a), and guilt feeling's regression coefficient in model 3 (b). This was calculated with the following formula, suggested by Kenny, Kashy, and Bolger (1998). In the formula, S_a = the standard error of a; S_b = the standard error of b:

$$Z = \frac{ab}{\sqrt{a^2 S_b^2 + b^2 S_a^2 + a^2 b^2}}$$

Calculated with the coefficients and their standard errors in the above regression models($a = 0.745$, $S_a = 0.023$, $b = 1.482$, $S_b = 0.568$), this yields the result: $Z = 2.60$, $p < .01$. Therefore, the partial mediating effect of guilt feeling was significant.

Discussion

In this study, the multidimensional conceptualization of guilt and the accordingly developed instrument, the BGS, was applied to explore answers to relevant questions that have attracted bereavement researchers' interest. To be specific, it explores the prevalence, background risk factors of bereavement guilt, and its relationship to complicated grief.

How common is guilt

Different types of guilt have different prevalence levels among bereaved adult children. This is consistent with the pattern of the discrepant findings from previous studies, which reported that from 7% to over 90% of bereaved people have experienced guilt in different periods of their grief. In general, the most commonly mentioned type of guilt, responsibility for death, was experienced by 37% of Chinese adult children. The frequency of this kind of guilt is in about the middle, when compared with other types. Guilt related to hurting the deceased was the lowest, while indebtedness guilt was the most prevalent, followed by guilt feelings.

Cultural influences may play a role in this situation. Indebtedness guilt implies that one feels obligated to repay another person who has previously provided the person great benefit. Paying back, in the context of receipt benefit (that is, reciprocation), is an important moral code guiding interpersonal relationships in the Chinese culture (Hwang, 1987). It is also one of the main themes in filial piety, which guides the interaction between parents and their adult children (Yeh, 2002). Reciprocation is a more significant and common concept in Chinese society than in western society (Rothbaum & Xu, 1995). Since paying one's parents back is a valued and prevalent moral code in Chinese society, any perceived failure to pay back tends to induce severe guilt. This may contribute to the high prevalence of indebtedness guilt in this sample of bereaved people.

It is worth noticing that there are a considerable number (70%) of bereaved adult children who felt great distress related to their bereavement guilt. This is the case even though the prevalence of certain types of guilt is not high; for example, hurting the deceased (23%) and survivor guilt (27%). This implies that any kind of guilt cognition can induce distressful guilt feelings.

The higher rate of feeling guilt in people bereaved due to an unnatural death compared to those who are bereaved through a natural death is consistent with previous findings (Miles & Demi, 1991 – 1992). This also confirms that the rate of guilt tends to be higher among certain vulnerable groups. Therefore, the question of how frequent bereavement guilt is, has no simple answer, because this depends on what types of guilt occur in what groups of bereaved people.

Risk factors

The risk factors explored in this study focus on demographic and contextual variables. Some factors have been investigated in previous studies, such as gender and religion of the bereaved, time since the death, cause of the death, and age of the deceased; while other factors, such as age, education level and marital status of the deceased, as well as gender of the deceased and psychological preparation for the death, were first explored in this study as potential risk factors for guilt in bereavement.

The finding that guilt in bereavement does not decrease with time is in accordance with some previous studies (Akiyama et al., 2010; Kowalski & Bondmass, 2008; Mizota et al., 2006; Zisook & Shuchter, 1985). Similar to the findings in one previous study (Gamino et al., 2000), younger ages of the deceased predicted higher guilt in family members. As for cause of death, similar results was found between the present study and some previous research studies, showing that unnatural death is associate with higher bereavement guilt, than is natural death (Gamino et al., 2000; Hazzard et al., 1992; Weinberg, 1994).

One advantage of the present study is that it investigates the predicting effects of all the risk factors simultaneously. After eliminating potential confounding variables, only three factors account for most types of guilt. None of the three factors has been previously documented in the literature on bereavement guilt. They are as follows: age of the bereaved, gender of the deceased parents, and psychological preparation for the death. This indicates that these three variables reliably predict total guilt in bereavement, as well as each sub-type of bereavement guilt (except for hurting guilt, which was not predicted by psychological preparation).

The negative correlation between age of the bereaved children and guilt intensity may relate to three aspects. The first regards attribution style. Younger adults are more likely to draw internal attributions, and older adults are inclined to draw external attribution (Blanchard-Fields, Baldi, & Stein, 1999). In evaluating responsibility for traumatic events, younger adults are more likely to attribute responsibility to the victim in the event (Daniel, 1985). In addition, younger adults reported more self-blame when encountering negative events than did older adults (Irion & Balanchard-Rields, 1987). It is possible that in facing death and related outcomes (such as inappropriate care for the parents), younger adults are more inclined to hold themselves responsible for the negative consequences (e. g. death of the parents; suffering of the parents), thus experiencing more guilt. Another possible reason involves the development of moral judgment. As developmental psychologists have suggested, moral judgment usually develops from outcome based to intention based (Armsby, 1971; Piaget, 1932). That is, in younger children, the "naughtiness" of a person who has done something wrong or caused a negative outcome depends mainly on how severe the negative outcome is. However, as age increases, the consideration given to the intention of that transgressor will be given more weight in moral judgment. If this developmental difference extends to adulthood, then it is understandable that younger participants in the present study felt more guilt than older participants: Younger adults think less about their innocent intentions or their ability to control the event. They naturally feel more guilt than do older adults, who take into consideration more comprehensive factors, and therefore feel less guilt. Third, it is a fact that younger adult children have had less time and ability to pay back their parents or to resolve possible regrets and conflicts in the relationship, compared to older children, considering that older adults are more mature and financially capable to take care of their aging parents.

Losing a mother is a risk factor for intense guilt. This may correlate with the severe distress associated with losing a mother. Guilt is an emotion that is closely linked more to consequences than to controllability (Graham, Doubleday, & Guarino, 1984). The more severe the outcome is, and the more negative emotions an individual experiences, the more guilt he/she feels. For example, one study found that the degree of negative emotions experienced by bereaved parents 3 weeks after the death predicted the level of counter – factual thinking (what one would have done to prevent the death) (Davis et al., 1995). Studies have found that losing a mother is generally more distressful and causes more negative influence on bereaved adults than losing a father (Horowitz et al., 1984). Following this line of reasoning, those who have lost their mothers are at higher risk of intense guilt.

Low psychological preparation is associated with higher guilt. This influence may be due to the fact that unexpected death is more difficult for the bereaved person to accept and integrate into their autobiographic memory, which in turn induces higher distress and pathological grief reactions. Another possibility is that when one is aware of the coming death of the parents, he or she may take more initiative to resolve conflicts and deal with unfinished business in the relationship, in order to reduce the chance of feeling guilt after the death. For example, the adult child may cherish the time spent with the parents, and take the chance to resolve the conflicts in their relationship, or do things that he or she feels obligated to do for the parents. In cases where the adult child was not aware of the impending death of the parents, he or she may think there is plenty time and chance in the future to fulfill his/her obligations or to make reconciliation. However, when the death happens suddenly, they feel more guilt than do those who have prepared for the death of their parents.

Previous studies have suggested that women experience more guilt in bereavement than men (Lang et al., 1996; Mizota et al., 2006). However, the present study finds that women only reported higher score than men on indebtedness guilt and guilt feelings, and not on other components of guilt or on total guilt. Previous studies which found gender difference were researching bereaved parents, but this study is carried out among bereaved adult children. Whether different relationships with the deceased account for the discrepancies in the result is up to further investigations.

Religion was reported as a protective factor for bereavement guilt (Bohannon, 1991b), but this result was not replicated in the present findings. Conversely, those with religion reported higher guilt in responsibility for the death, but not in other types of guilt. The different types of religion (Christian in Bohannon's study, compared with a mixed group of Christian, Buddhist, and Taoist in the present study), and the difference between religious practice and religious belief (Bohannon's study recorded the frequency of going to church, while the present study asked about participants' religious beliefs) may play a role in explaining the contradictory findings.

The multivariate analysis found that some risk factors only predict certain types of guilt and not others. For example, gender only predicts indebtedness guilt and guilt feelings, while unnatural death induces higher indebtedness guilt, but not other types. People with religious beliefs were at a higher risk to experience guilt related to responsibility for the death, but not regarding other kinds. Such results show that different types of guilt are associated with different predictors. It confirms the advantage of the BGS in delineating guilt into sub - components.

Guilt and complicated grief

Consistent with some previous studies (Boelen & Lensvelt - Mulders, 2005; Boelen et al., 2003), a positive relationship between guilt and complicated grief is found in the present study. In addition, the present study also finds that guilt can explain additional variance in complicated grief beyond demographic variables, death related variables and psychological distress. A similar effect of guilt on complicate grief was also reported by a previous study on suicide bereavement (Fischer, 2006). These results extend the understanding of the relationship between guilt and complicated grief and establish a closer relationship between these two concepts.

In addition to the analysis on the total guilt level, more analysis has been done to explore different types of guilt and complicated grief. Each of the sub – components of guilt was found to be significantly associated with symptomatic grief reactions. In addition, the association between each guilt component remained significant even when demographic variables, death related variables, and depression and anxiety were controlled in the regression model. Most previous studies focused on the detrimental role of self – blame for the death, whereas this study demonstrates that other components of guilt are also associated with psychopathology. One recent study investigating a small group of clinical samples suggested that perceived responsibility for the death contributes to depression but not to complicated grief (Golden & Dalgleish, 2012). This is different from the present study, which recruited a large sample not restricted to people who fit the diagnostic criteria of complicated grief.

After the shared variances in guilt components were controlled for, survivor guilt and indebtedness guilt showed additional contribution to complicated grief in the regression model. These two types of guilt received little investigation in relation to complicated grief in existing literature. However, the results suggest that they most strongly associated with grief symptoms, and that they deserve more attention.

It is worth noting that the regression coefficient of guilt is greater than that of any demographic variable, death related variable, depression or anxiety. This strong association should not be attributed to the similarity of scale items, because the only guilt item in the ICG was removed from the analysis.

In the conceptual level, the relationship between guilt and complicated grief can be understood from the Cognitive – behavioral model of complicated grief (Boelen, van den Hout, et al., 2006). This model postulates that three core cognitive processes, including poor integration of the death into autobiographic memory, negative global beliefs and misinterpretations of grief reactions, along with anxiety and depression avoidance strategies, are responsible for the occurrence and persistence of complicated grief symptoms.

Guilt in bereavement implies difficulty in accepting and making sense of the death and its implication to one's life. As Parkes observed on a group of widows; "In agonizing over events which were often quite trivial they seemed to be looking for a chance to castigate themselves, as if by accepting blame they could somehow reverse the course of events and get back the missing spouse" (Parkes, 1996, p. 84). The interpersonal function of guilt (Baumeister et al., 1994) gives the bereaved person the sense of maintaining the relationship through unfinished business, suggesting difficulty in accepting that the death is unchangeable and that the relationship with the deceased person cannot be recovered, to be as it was before the death. Studies have found that the association between self – blame and continuing bonds implies poor integration of the death into one's present life, resulting in such actions as keeping the deceased's belongings (Field & Bonanno, 2001). Therefore, the relationship between guilt and complicated grief may be understood from the viewpoint of poor integration of the death into existing autobiographical knowledge.

The second process underlining complicated grief is negative global beliefs, including negative belief in oneself. Perceiving that one has failed one's own inner standards and expectations, or has contributed to awful consequences, naturally threatens self perception, and makes the person feel worthless and demonstrate low self – esteem. The link between guilt and negative belief in self is evident in previous literature (Boelen & Lensvelt – Mulders, 2005). The negative correlation between guilt and self – esteem is duplicated in the present data, as reported in Chapter 5. The mediating effect of self – esteem between guilt and complicated grief should be explored in future research.

In addition, guilt is also correlated with anxiety and depression avoidance.

Anxiety avoidance represents avoiding confronting and processing the memories related to the facts of the death. It is possible that processing the information of the death and past relationship will activate the person's guilt cognition, which is then accompanied by distressful feelings. Therefore people back off from approaching death related memories and information. Depression avoidance represents mourners' withdrawal and refraining from social, occupational and recreational activities. Blocking those activities also blocks positive reinforcements and emotions that can be helpful in adjusting to the loss. This pattern of behavior is linked to survivor guilt, in which mourners believe they do not deserve to live and enjoy themselves in a world without the deceased. Holding this belief will lead them to activate depression avoidance strategy. The mediating effect of avoidance coping strategy between guilt and negative psychological consequences has also been suggested by previous studies (Boelen, van den Bout, & van den Hout, 2006; Held et al., 2011; Street et al., 2005).

In addition to the above analysis with the Cognitive – Behavioral model of complicated grief, another postulation about the contribution of guilt to complicated grief is that self – blame in guilt adds extra distress on the bereaved person (Horowitz et al., 1984). The present findings are in accordance with this hypoproject, as guilt distress was found to mediate the association between guilty thoughts and complicated grief symptoms. However, guilt distress did not fully mediate the contribution of guilt cognitions on complicated grief, suggesting the possibility of other mechanisms which underline the relationship. For example, guilt may impair one's perception of self value, which in turn contributes to psychopathology (Weinberg, 1994); guilt in bereavement may be correlated with maladaptive continuing attachment, which is claimed to be responsible for hindering adjustment (Field & Bonanno, 2001; Field et al., 2000); or guilt may be correlated with maladaptive strategies (such as avoidance coping), thereby impeding adjustment (Boelen & Lensvelt – Mulders, 2005). However, these propositions have been insufficiently tested or supported by research data, and more empirical investigation in the future is needed to uncover the mechanisms behind the association between bereavement guilt and complicated grief. The present study has stepped out toward this direction, and has found that extra distress in guilt mediates the influence of self blame. The influence of negative cognitions on complicated grief has been documented in the literature (Boelen & Lensvelt – Mulders, 2005; Fleming & Robinson, 2001). The mediating effect of guilt feelings supported the statement that at least some of the influence of the negative cognitions, such as guilt, comes about via the distress accompanied by those negative thoughts.

Limitations

Several strategies were adopted to ensure research rigor; such as a large sample size with a diverse background of participants in terms of demographics (e. g. age, gender, religion, education level, marital status) and death related conditions (e. g. bereavement time and cause of death). However, there are some limitations of this research that need to be pointed out.

First, it should be noted that the cross – sectional design of this study does not draw a causal relationship between guilt and complicated grief. Even though there are theories to support the influence of negative cognitions on emotional problems and difficulties in adjusting to grief (Boelen, van den Hout, et al., 2006), it is presumptuous to state that guilt causes complicated grief based merely on the findings from this study.

Second, there are three groups of risk factors in the framework proposed by W. Stroebe and Schut (2001); situational, personal and interpersonal. In this study, only parts of the first two types of risk factors were explored. The demographic and death related risk factors identified in this study were helpful in identifying an at risk group of bereaved people, as these factors existed before or at the moment of the death. However, these factors cannot be changed (since they are factual variables), and therefore have little application in treating bereavement guilt. It is difficult for one study to explore all three types of risk factors, and further studies are needed to compensate for this limitation.

Third, as discussed in Chapter 5, there is no evidence to state that the on-line sampling is representative of the general population of Chinese adult children. Therefore caution is needed when trying to draw a generalized conclusion.

Summary

In summary, the present study has answered the research questions raised at the beginning of this chapter. Specifically, it has found that:

In general, about 40% of bereaved Chinese adult children have experienced significant levels of guilt, but that the prevalence of different types of guilt ranges from 23% to 81%. Guilt is more commonly experienced by bereaved people who lost parents due to an unnatural death than to a natural death. Indebtedness guilt was the most common among both natural death and unnatural death groups.

Three core predictors (age of the adult child, age of the deceased parents, and psychological preparation) reliably predict total guilt, as well as most subtypes of guilt in bereavement.

Bereaved adult daughters experience higher indebtedness guilt and guilt distress than do bereaved adult sons. Those with religious beliefs reported higher guilt in guilt about responsibility for the death, and those bereaved through a traumatic death experience higher guilt related to indebtedness guilt.

The level of total guilt is significantly associated with complicated grief.

The level of total guilt predicts complicated grief over the influence of demographic and death related variables, and degree of psychological distress.

Each guilt component is positively correlated with complicated grief, and each of them show additional contributions to complicated grief even when demographic variables, death related variables, and psychological distress are controlled.

Survivor guilt and indebtedness guilt have additional contributions to complicated grief, beyond other types of bereavement guilt.

The association between guilt cognitions and complicated grief is partially mediated by guilt feelings.

Chapter 7

General discussion and conclusion

Summary of the research and findings

This project was inspired by noticing the lack of attention toward bereaved adult children's grief experiences. Therefore it focuses on this subgroup of bereaved people. Meanwhile, the emerging trend to explore and gain more understanding related to each emotion in grief and the acknowledged importance and prevalence of guilt in grief are key background features informing this research project. Despite the significant guilt in bereavement, the following gaps were identified in the literature:

A thorough review on guilt in bereavement reveals that this phenomenon has not been properly defined or measured. Most previous studies focus on guilt over responsibility for the death, ignoring other possible content in guilt that have been suggested by certain qualitative inquires. Meanwhile, the affective component of guilt has usually been overlooked. In most previous studies, only the cognitive aspects of guilt were measured, despite the fact that self – blame thoughts are usually accompanied by uncomfortable feelings.

There is lack of convergence on the conceptualization and measurement of guilt in general psychology. Therefore it's necessary to develop a domain specific definition and measure for guilt in bereavement.

Researchers have made efforts to understand the prevalence and predictors of guilt in bereavement, as well as its association with grief outcomes. However, due to the deficiencies in conceptualizing and measuring guilt, previous studies yield mixed findings regarding these three themes. Specifically, there is huge range recorded for the prevalence of guilt in bereavement, probably due to different studies investigating differing types of guilt within varying subgroups. The risk factors explored before were fragmented, and problems remain as they did not use multivariate analysis to control confounding variables. The relationship between guilt and health outcomes concerned mainly depression and anxiety, while few studies investigated complicated grief, which is a distinct group of pathological reactions in bereavement.

Given the gaps and limitations in previous studies, this project has aimed to achieve the following objectives:

Establish a comprehensive conceptualization of bereavement guilt, which is expected to be a multidimensional construct.

Explore the structure of and components found within bereavement guilt.

Develop and validate a tool which is in accord with the conceptualization of bereavement guilt, and which has good psychometric properties.

Apply the newly developed conceptualization and measure to address mixed findings or limited findings in previous studies. Specifically, answer the following questions: How frequent is guilt in bereaved Chinese adult children? What demographic and death related factors predict higher guilt in bereaved Chinese adult children? What is the relationship between bereavement guilt and complicated grief?

Combining both in – depth interviews and a questionnaire survey, all the above mentioned research objectives have been achieved. Specifically, this study has yielded the following results:

Bereavement guilt is conceptualized as a multidimensional construct. It contains both cognitive and affective aspects with five factors: responsibility for the death, hurting the deceased, survivor guilt and indebtedness guilt and guilt feelings.

The newly developed Bereavement Guilt Scale (BGS) comprises 14 items, which load on 5 factors of bereavement guilt, as mentioned above. It is proven to obtain good dimensionality, as well as construct validity. It was also proven to have good internal consistency and test – retest reliability. Since the BGS reflects the complexity of bereavement guilt, and delineates this construct into sub – components, it is suitable for adoption in exploring an in – depth understanding of guilt in bereavement.

The prevalence of guilt ranges from 23% to 81% in general bereaved adult children depending on the specific type of guilt. Guilt is more common among the natural death group than among the unnatural death group.

In controlling confounding variables, three core predictors (i. e. younger age of the bereaved child, losing a mother, and low psychological preparation for the death) have been identified to predict high scores on most types of guilt. Meanwhile, different types of guilt were associated with different demographic and death – related predictors. For example, the gender of the bereaved child only predicts indebtedness guilt and guilt feelings; religious beliefs only predict guilt about responsibility for the death; and an unnatural death only predicts indebtedness guilt, when other factors are controlled.

It has extended the relationship between bereavement guilt and depression to pathological grief symptoms. It has found that not only the total guilt score, but also the score for each component of guilt is positively associated with complicated grief. Bereavement guilt shows additional association with complicated grief when background factors, depression and anxiety, are controlled. The strong relationship between guilt and complicated grief further confirms the significance of this emotion in the bereavement experience.

This project has filled the gaps in defining and measuring bereavement guilt, and illustrates the application of the newly proposed conceptualization and measure toward exploring scientific understanding of bereavement guilt. Its contributions and applications will be further discussed below.

Contributions and implications

Theoretical aspects

The multidimensional conceptualization of bereavement guilt

Responding to the call for developing a domain specific definition and measurement of guilt, this project has constructed a multidimensional conceptualization of bereavement. As a result, bereavement guilt is defined as:

"A remorseful emotional reaction in grieving, with the recognition of having failed to live up to one's own inner standards and expectations in the relationship to the deceased, or of being responsible for unwanted consequences."

This definition comprises both cognitive and affective components. There are multiple sources for the cognitive aspect. Therefore, in this conceptualization bereavement guilt is a multidimensional construct. This conceptualization has several features:

First, it follows the historical treatment of guilt in bereavement; that guilt has been considered an emotional element in grief, and has been incorporated into grief measurement accordingly. Second, it integrates key elements in previous definitions of guilt, comprising cognitive evaluation, inner standards, responsibility, and uncomfortable feelings. Third, it is in accordance with the actual experiences of bereaved people, which were collected and analyzed in study one. Fourth, the multidimensional structure of bereavement guilt has been confirmed by empirical data through Confirmatory Factor Analysis in study two. Finally, previous conceptualizations of guilt in bereavement can be incorporated into the present frame. For example, self – blame for the death, which was mostly examined in bereavement studies related to bereavement guilt, can be seen as one source within the aspects of guilt cognition.

The multidimensional conceptualization of bereavement guilt runs parallel to theories and findings in other domains. For example, trauma – related guilt comprises several different components of cognitive dimension and one guilt distress dimension. (Kubany & Abueg, 1996). Another case in point is care giver guilt, which is also a domain specific multidimensional construct (Losada et al., 2010). This conceptualization of bereavement may provide a common starting point for bereavement researchers, who can increase their efforts in exploring more understanding of guilt in bereavement, and its role in physical and psychological adjustment.

The necessity and advantage of delineating guilt into sub – components is also demonstrated from the quantitative understanding of its characteristics and relationship to outcomes. First, it is found that different types of bereavement guilt have different prevalence, and the discrepancies were considerable. To some degree it helps explain the widely varying range of reported bereavement guilt prevalence in previous literature. Second, despite the fact that various types of guilt share several common risk factors (e. g. age of the bereaved), there are distinct predictors to each respective type of guilt. The predicting effect of some risk factors on certain types of guilt but not others may account for the discrepant findings in previous literature that investigated risk factors for guilt. For example, an unnatural death does not necessarily induce higher guilt on responsibility for the death, but it does predict more intensive indebtedness guilt. Third, the association between each component in bereavement guilt and health outcomes have been assessed and compared. The findings suggest that even though all types of bereavement guilt are positively correlated with complicated grief, survivor guilt and indebtedness guilt show stronger and additional connections to complicated grief. This result has identified two core types of guilt that may be extremely detrimental in adjusting to loss, and that need special attention and intervention. The findings also suggest different mechanisms behind the associations connecting each type of guilt to the health outcomes. Those issues are out of the scope of this project but can be explored in a future study.

Extended understanding of the association between guilt and health outcomes

The role of guilt in psychological adjustment has long been evaluated. Despite some inconsistent findings, many previous studies have documented the association between guilt and psychopathology, such as depression (Gupta et al., 1988), anxiety (Lowe, 1964), OCD (Shapiro & Stewart, 2011) and even PTSD (Beck et al., 2011; Henning & Frueh, 1997). However, due to the limited number of studies concerning the role of guilt in bereavement, little evidence about the relationship between guilt and complicated grief exists in previous literature, with just a few exceptions. The findings of this project have contributed to the empirical evidence of the relationship between guilt and psychopathology in extending the outcomes from psychological distress to specific grief symptoms. It demonstrates a strong connection between bereavement guilt (as well as each component of guilt) and complicated grief, even when background variables and psychological distress were controlled. So far, the mechanism underlining such a relationship is unclear, but a preliminary test of the mediating role of guilt feelings has been conducted in this project, and proves that at least some part of the influence of guilt cognitions on complicated grief can be attributed to the distressful feelings accompanying those thoughts of guilt.

The model of risk factors, bereavement guilt and health outcomes in bereavement

Taken together, the main findings of this project can be presented as in Figure 7. 1. These findings indicate the multidimensional conceptualization of bereavement guilt, the prevalence of each type of guilt, which risk factors predict higher guilt, and the strong associations between guilt and negative health outcomes in bereavement. This model has not been directly tested by Structural Educational Modeling, but the structure of bereavement guilt and the relationship among those constructs has been established by separate statistical analysis.

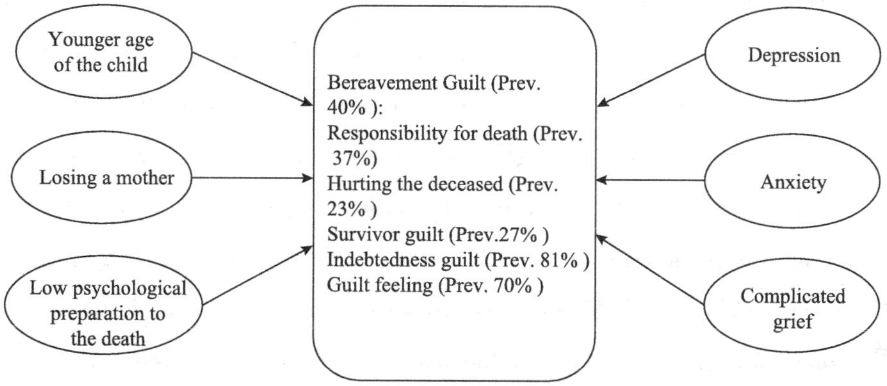

Figure 7. 1 Risk factors, Bereavement Guilt and Health Outcomes

Prev. : Prevalence of total/subtypes of guilt.

Improved understanding of the grief experience of Chinese bereaved adult children

This is the first study, to my knowledge, concerning the guilt experience among Chinese adult children, which is a relatively ignored group of bereaved people. Bereavement studies focusing on adult children are less than among other population groups, such as spouses and parents, despite the enormous number of people included, and the significant influence of the death on them. Moreover, the moral codes in Chinese society, which emphasize duties and obligations, as well as the cultural beliefs of filial piety concerning the obligation of adult children to their parents, adds extra importance to exploring the guilt experience of this group of people. Some cultural beliefs permeate the bereaved adult children's narratives about their guilt. For example, staying with the dying parents and paying their last respects at the moment of death is a Chinese filial tradition. Those who failed to do so reported self – blame for not being able to fulfill this culturally prescribed obligation as a daughter or son.

This project indicates that guilt may be a common experience within this population group, as some types of guilt can be as high as 81%. For most of them, the guilt experience is disturbing, because about 70% of them reported prominent distressful guilt feelings. From the narratives in study one, it can be seen that intensive guilt is an extremely negative experience, judging from the words of one participant: "... *as if you keep stabbing and thrusting yourself* [*through*] *with a knife*". These evidences suggest that this emotion in this group of people deserves more attention and understanding from both researchers and clinicians. The present project has contributed to the knowledge about guilt in bereavement, bereaved adult children, and Chinese bereaved people's experiences.

An effort to understand specific emotions in grief

There are all kinds of emotions that the bereaved person may experience within their bereavement, and *grief* refers to the combination of all kinds of emotions. Researchers have now established the idea that not all negative emotions experienced by bereaved people are pathological, but one group of reactions demonstrate the problematic grief reaction, which is called complicated grief. Some emotional reactions in grief have been included as indicators of complicated grief; such as excessive bitterness and anger. It is worth noting that some emotions in grief are not included in complicated grief, and these have not attracted thorough investigation. Guilt, a long recognized emotion, is one of them. The nature and relationship of guilt to complicated grief has not previously been clear, given the definition ambiguity and measurement difficulty. This project is an effort to stress the complexity of grief and establish a better understanding of the specific emotion of guilt in grief. Guilt is not a part of complicated grief, but it may play an important role in adjusting to loss. With the development of bereavement research, it is expected that one day there will be a systematic understanding of each emotion in grief, along with the nature of each and the roles each played in the negative consequences of grief (such as depression or complicated grief), or in positive outcomes (such as personal growth). This project, focusing on guilt, is a step in the right direction.

Methodological aspects

Mixed study design

Methodological pluralism has been advocated by experts in the bereavement fields, along with the advantage in combing both qualitative and quantitative research methods to develop bereavement related scales (Robert Neimeyer & Hogan, 2001). This project is a mixed method program, which sequentially adopts qualitative and quantitative research methods. The different types of data collected (both narratives and numbers) in this research program add to the richness of the findings and enriches the understanding of the guilt experienced by bereaved adult children from different perspectives. It is one of a few bereavement studies which have adopted a mixed method design to investigate a phenomenon in bereavement (Robert Neimeyer & Hogan, 2001).

A tool for future research: The Bereavement Guilt Scale (BGS)

One direct product from this project is the development of the BGS. This scale has been validated and proven to have good content validity, factorability, construct validity, internal consistency and test – retest reliability. This is the first scale that has been developed from bereaved people by which to test guilt. This scale also reflects the complexity of bereavement guilt and delineates it into different components. It enables the assessment of the nature and consequences of not only general guilt in bereavement but also of each of its aspects. As stated above, it can be used to address the discrepant findings within previous literature and provides a refined answer to those questions that have not been satisfactorily answered before.

Practical implications

Awareness of all kinds of guilt

Responsibility for the death was the kind of guilt most discussed in previous literature on guilt in bereavement. The findings of this project call more awareness to all kinds of bereavement guilt, apart from just responsibility for the death. It should be noted that some types of guilt are more prominent and common among bereaved adult children than is responsibility for the death; such as indebtedness guilt. Moreover, all types of guilt are accompanied with distressful guilt feelings, and they are all associated with complicated grief. Therefore, the indicators of guilt should not be restricted to this one indicator alone. The manifestation of any type of guilt deserves attention from friends or family members of bereaved people, as well as from clinical professionals, given that bereavement guilt is associated with the symptoms of complicated grief.

Identify at – risk groups of bereaved adult children

The results of this study suggest that some bereaved adult children at greater risk of experiencing guilt than are others. Those risk factors identified can help supporting professionals to detect vulnerable people, and to be ready to provide corresponding education, prevention or intervention to those who need it.

One vulnerable population is that of bereaved college students, as younger ages are a risk factor for higher bereavement guilt. The bereavement rate in college students is not low, as a recent survey found that 30% of college students have experienced the death of a loved one within the past 12 months, and 39% of these students had experienced this loss within the preceding 24 months (Balk, Walker, & Baker, 2010). Most college students are young adults. Apart from the pain and distress of losing a loved one through death, this group is more likely to experience self – blame and self reproach stemming from bereavement guilt. In fact, according to another study we carried out recently, the sense of loss and guilt were rated as the two most distressful emotions, by a group of bereaved college students (J Li, Chan, & Chow, 2011, November). Therefore clinical professionals who work in college should be aware of the risk to bereaved college students who have lost their parents, especially to those who lost their mothers, as the gender of the deceased parents is also a risk factor.

Similarly, higher risk would be experienced by those bereaved through unnatural causes; such as suicide, accident, homicide and natural disaster, or those bereaved through a natural death but who had little psychological preparation. Professionals who work with these subgroups of bereaved people should be sensitive to this populations' vulnerability, and provide adequate support to address and deal with their guilt.

Dealing with bereavement guilt

The high prevalence of bereavement guilt and positive association with complicated grief symptoms endorses the need for clinical professionals to pay attention to this grief emotion. As the data suggests, about 70% of bereaved adult children report prominent guilt distress. Excessive guilt is also associated with greater risk of pathological grief reactions.

Some initiatives can be taken prior to the loved one's death to reduce the risk of feeling guilt in bereavement. Some studies have suggested the importance of a "good ending" in reducing the risk of guilt. For example, by helping the family to resolve conflicts in the relationship before the death, or by carrying out end of life discussions with the deceased about the attitude toward death or after death arrangements are possible strategies to reduce the risk (Jonasson et al., 2011). Medical professionals can also try optimized strategies for breaking the bad news to family members, for the sake of their improved psychological preparation.

Guilt is difficult to take away. Simply telling the person, "You should not feel like that" does not usually help (Parkes, 2008). Rando (1993) has suggested the necessity of understanding possible functions of bereaved persons' guilt before trying to remove that individual's guilt. In some cases, the death is so devastating that it upsets the original meaning system of the bereaved person, causing them to lose their sense of control. In order to gain back a sense of control, the person may attribute the death to himself/herself, holding oneself responsible for the death (Janoff - Bulman, 1979). In some cases, guilt represents unfinished business within the relationship ("I once hurt you and I owe you apology", or "I am indebted to you because I did not pay you back for the benefit you gave to me"). Therefore, guilt may function as a way to maintain a continuing relationship with the deceased. Studies have found that the association between self - blame and maladaptive continuing bonds (Field & Bonanno, 2001; Field et al., 2000), as well as survivor guilt, is considered to be a means of identifying with the deceased person (Kolb, 1983). If guilt bears with it these functions, the bereaved person will not let go of their guilt easily. In such cases, alternative and adaptive ways to maintain those functions and satisfy the bereaved person's needs should be explored in order to help the person. For example, if the guilt functions as a continuing bond, then other adaptive approaches that can help the person maintain that bond should be explored.

Another inspiration stimulated from the findings is the core position of "responsible for negative consequences" within bereavement guilt. Participants in the interview study have revealed that perceived contributions to negative consequences underline their experience of guilt. Guilt is based on the sense that they have contributed to the death, imposed harm on the parents, or deprived the parents of due benefits (e. g. failed to pay the parents back). The subsequent factor analysis confirmed that self attribution of these consequences is main factors of bereavement guilt. It seems that extreme internal attribution of negative outcomes underlines bereavement guilt. Therefore, when working with bereaved persons with excessive guilt, a possible approach is to help the person reassesses his or her responsibility related to the outcome: Was he/she fully accountable for the outcome? Are there other factors that play a role? As discussed in Chapter 6, younger adults are at higher risk of bereavement guilt than are older adults, probably due to younger individuals being more likely to focus on the outcomes but oversee other factors (e. g. intention of the person). This suggests applying a cognitive approach to help the bereaved person gain a rational perception of their responsibility in the event, and accept their limitations when things have gone wrong or gone beyond their intentions. The effect of cognitive therapy on guilt has been reported by some researchers. For example, one study found that cognitive therapy can reduce the guilt felt by suicide survivors (De Groot et al., 2007). Cognitive therapy has also been proven to be effective in treating trauma related guilt (Kubany & Manke, 1995).

In a word, guilt in bereavement is a distressful emotion that should not be ignored, and more efforts are needed to prevent or alleviate its risk and negative consequences.

Limitations

The limitations of each study in this project project have been discussed in their respective chapters. In addition to this, there are several general comments that will be outlined next.

Shared sample for scale validation and model testing

In the ideal scenario, the scale should be validated first before applying against a large group of participants for further exploration of the relationship between relevant variables. However, due to the limits of time and resources, this project collected data for the scale validation and model testing at the same time. This strategy was based on the consideration of the difficulties in recruiting a large number of bereaved participants. The result is that participants had to fill in some items which were not used in the model testing period, because they were deleted during the validation process. However, no extra distress has been imposed on the participants, and the relationships tested in the scale validation process did not duplicate those in the model testing process.

Not differentiating legitimate from illegitimate guilt

It is very difficult to assess to what degree a participant's guilt is based on fact or simply a subjective delusion. For example, if the participant claim responsibility for the death, there is nowhere in the data can provide factual information so that an objective evaluation can be made to judge how much responsibility should be actually be attributed to the person. Some bereavement experts have suggested differentiating legitimate from illegitimate guilt before treating guilt in bereavement (Rando, 1993). However, in the present conceptualization, guilt is considered as a subjective experience: whether this feeling or evaluation was legitimate or illegitimate is out of the scope of this research. This research is concerned only with subjective cognitions and feelings. Whether legitimate and illegitimate guilt associates differently with grief outcomes could be a direction for future research.

Restriction to a specific subgroup

This project aims to explore the guilt experience among bereaved Chinese adult children. Therefore participants in all three studies were restricted to this sub-group of bereaved people. As role obligations and expectations may be different across different relationships, the nature and consequences of bereavement guilt may be different among bereaved adult children, spouses and parents. More studies conducted among other subgroups of bereaved people, or within a mixed population, are needed before generalizing the findings of this project to bereaved people as a whole.

Embedded cultural context

As mentioned before, cross-cultural differences are also evident in the grief experience, along with guilt content. The present study is embedded in the Chinese cultural context, and also reflects the influence of cultural beliefs. For example, the high prevalence of indebtedness guilt in this study may reflect the cultural beliefs of reciprocation and filial piety. Different patterns of findings are expected from studies conducted in western countries. Also, the characteristics of participants require caution when interpreting some results. For example, most Chinese people have no religious beliefs, even though most of them are influenced by Buddhism and Taosim. One of the risk factors, religious belief (as mentioned in this project), comprises mainly Buddhism and Taosim. However, if the study were to be carried out in a country with a higher prevalence of Christianity, the result might be different. In conclusion, validating the BGS in western countries and applying this tool in other cultures is necessary to gain a greater understanding of bereavement guilt.

Future directions

Figure 7.2 is the proposed structure model which may provide directions in which to investigate bereavement guilt. The first step in this model is to conceptualize bereavement guilt and develop a valid measure by which to capture the conceptualization. Then each pathway in the model can be explored, including risk factors for bereavement guilt, the relationship between guilt and health outcomes in bereavement, any mediators that transfer association, and any moderator that may influence the direction or magnitude of the association.

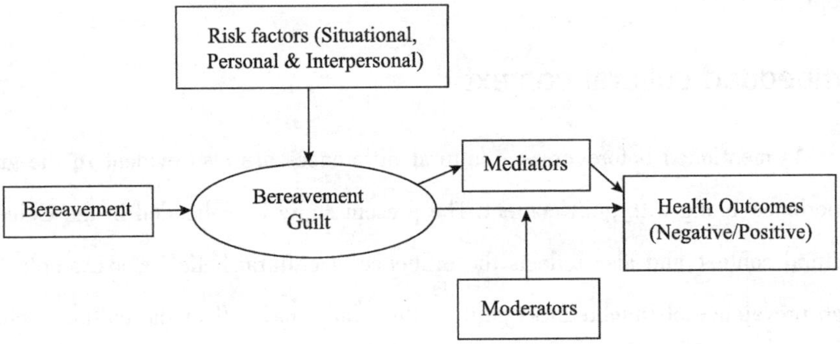

Figure 7.2 Structural Model for Investigating Bereavement Guilt

This project has accomplished both conceptualization and the development of a measurement. It has explored the demographic and death – related predictors related to bereavement guilt. It has also established the association between bereavement guilt and complicated grief. However, the scope of this project research is limited. More empirical studies are needed to generalize the findings of the present project and to move the exploration forward, as indicated in the structure model.

First, since the present conceptualization, scale and other empirical findings are based on data collected from Chinese bereaved adult children, it is necessary to confirm the conceptualization, validate the scale, and see if the understandings regarding bereavement guilt in adult children are replicated among other bereaved subgroups, or among mixed groups of bereaved people.

Second, the present study has explored demographic and death – related risk factors. The findings are helpful in identifying which groups of people are most at risk, but less useful in informing guilt intervention, because those factors are unchangeable. Therefore, future studies are encouraged to explore alterable risk factors that can be applied in treating bereavement guilt.

Third, outcomes in bereavement include not only negative aspects but also positive aspects, such as personal growth. So far, evidence has been documented on the positive relationship of guilt to depression, anxiety and complicated grief. Future studies may extend this direction of exploration to assess the relationship between guilt and positive outcomes in bereavement. It is also intriguing to explore under what conditions or for whom guilt might play an adaptive role in psychological well – being.

Fourth, as shown in the structural model, potential mediators, such as a sense of worthlessness, maladaptive coping strategies and continuing bonds, that may explain the mechanism of the association and the moderators that can alter the relationships are both meaningful ways for future exploration.

Fifth, the cross - sectional design of the present study cannot draw a causal conclusion between guilt and complicated grief. Longitudinal studies conducted in the future to address the predicting effect of guilt on health outcomes will provide more evidence of the role of bereavement guilt, as well as the etiology of positive or negative changes in bereavement.

Sixth, cross - cultural differences are also a direction that deserves future exploration. Studies conducted in western countries can test whether the present conceptual model based on the Chinese population is also applicable within the western population. Findings in cross - cultural studies will shed light on the understanding of the nature and influence of bereavement guilt in varying cultural contexts.

In conclusion, this project has proposed a common starting point for studies on bereavement guilt and outlined a model on which other researchers may take the combined initiative to move our understanding of the nature and influence of bereavement forward. The ultimate aim of such initiatives is to improve our knowledge in this field and provide better support and treatment of complications in grieving associated with extreme guilt feelings.

Reference

Adolfsson, A., & Larsson, P. G. (2004). Guilt and emptiness: Women's experiences of miscarriage. *Health Care for Women International*, 25, 543 – 560.

Akiyama, A., Numata, K., & Mikami, H. (2010). Importance of end – of – life support to minimize caregiver's regret during bereavement of the elderly for better subsequent adaptation to bereavement. *Archives of Gerontology and Geriatrics*, 50(2), 175 – 178.

Alexander, B., Brewin, C. R., Vearnals, S., Wolff, G., & Leff, J. (1999). An investigation of shame and guilt in a depressed sample. *British Journal of Medical Psychology*, 72(3), 323 – 338.

American Psychitric Association. (1994). *Diagnostic and Statistical Manual of Mental Disorders: DSM – IV.* Washington: American Psychiatric Association.

Anolli, L., & Pascucci, P. (2005). Guilt and guilt – proneness, shame and shame – proneness in Indian and Italian young adults. *Personality and Individual Differences*, 39(4), 763 – 773.

Armsby, R. E. A. (1971). A reexamination of the development of moral judgments in children. *Child Development*, 42, 1241 – 1248.

Bailley, S. E., Kral, M. J., & Dunham, K. (1999). Survivors of suicide do grieve differently: Empirical support for a common sense proposition. *Suicide and Life – Threatening Behavior*, 29(3), 256 – 271.

Balk, D. E., Walker, A. C., & Baker, A. (2010). Prevalence and severity of college student bereavement examined in a randomly selected sample. *Death Studies*, 34(5), 459 – 468.

Baron, R. M., & Kenny, D. A. (1986). The moderator – mediator variable distinction in social psychological research: Conceptual, strategic and statistical considerations. *Journal of Personality and Social Psychology*, 51(6), 1173 – 1182.

Barrett, K. C. (1995). A functionalist approach to shame and guilt. In J. P. Tangney & K. W.

Fischer(Eds.),*Self – conscious emotions: The psychology of shame, guilt, embarrassment, and pride*(pp. 25 – 63). New York: The Guilford Press.

Baumeister, R. F., Stillwell, A. M., & Heatherton, T. F. (1994). Guilt: An interpersonal approach. *Psychological Bulletin*, 115(2), 243 – 267.

Bear, G. G., Uribe – Zarain, X., Manning, M. A., & Shiomi, K. (2009). Shame, guilt, blaming, and anger: Differences between children in Japan and the US. *Motivation and Emotion*, 33(3), 229 – 238.

Beck, J. G., McNiff, J., Clapp, J. D., Olsen, S. A., Avery, M. L., & Hagewood, J. H. (2011). Exploring negative emotion in women experiencing intimate partner violence: Shame, guilt, and PTSD. *Behavior Therapy*, 42(4), 740 – 750.

Bedford, O. (2004). The individual experience of guilt and shame in Chinese culture. *Culture Psychology*, 10(1), 29 – 52.

Bedford, O., & Hwang, K. K. (2003). Guilt and shame in Chinese culture: A cross – cultural framework from the perspective of morality and identity. *Journal for the Theory of Social Behavior*, 33(2), 127 – 145.

Bentler, P. M. (1990). Comparative fit indexes in structural models. *Psychological Bulletin*, 107, 238 – 246.

Berman, L. E. A. (1978). Sibling loss as an organizer of unconscious guilt: A case study. *Psychoanalytic Quarterly*, 47(4), 568 – 587.

Berrios, G. E., Bulbena, A., Bakshi, N., & Dening, T. R. (1992). Feelings of guilt in major depression: Conceptual and psychometric aspects. *The British Journal of Psychiatry*, 160, 781 – 787.

Bjelland, I., Dahi, A. A., Haug, T. T., & Neckelmann, D. (2002). The validity of the Hospital Anxiety and Depression Scale: An updated literature review. *Journal of Psychosomatic Research*, 52(2), 69 – 77.

Blanchard – Fields, F., Baldi, R., & Stein, R. (1999). Age relevance and context effects on attributions across the adult lifespan. *International Journal of Behavioral Development*, 23(3), 665 – 683.

Boelen, P. A., & Lensvelt – Mulders, G. J. L. M. (2005). Psychometric properties of the Grief

Cognitions Questionnaire(GCQ). *Journal of Psychopathology and Behavioral Assessment*, 27 (4), 291 - 304.

Boelen, P. A., & van den Bout, J. (2005). Complicated grief, depression, and anxiety as distinct postloss syndromes: A confirmatory factor analysis study. *American Journal of Psychiatry*, 162, 2175 - 2177.

Boelen, P. A., van den Bout, J., & van den Hout, M. A. (2003). The role of cognitive variables in psychological functioning after the death of a first degree relative. *Behavior Research and Therapy*, 41(10), 1123 - 1136.

Boelen, P. A., van den Bout, J., & van den Hout, M. A. (2006). Negative cognitions and avoidance in emotional problems after bereavement: A prospective study. *Behavior Research and Therapy*, 44(11), 1657 - 1672.

Boelen, P. A., van den Hout, M. A., & van den Bout, J. (2006). A cognitive - behavioral conceptualization of complicated grief. *Clinical Psychology: Science and Practice*, 13(2), 109 - 129.

Bohannon, J. R. (1991a). Grief responses of spouses following the death of a child: A longitudinal study. *Omega: Journal of Death and Dying*, 22(2), 109 - 121.

Bohannon, J. R. (1991b). Religiosity related to grief levels of bereaved mothers and fathers. *Omega: Journal of Death and Dying*, 23(2), 153 - 159.

Bonanno, G. A., Neria, Y., Mancini, A., Coifman, K. G., Litz, B., & Insel, B. (2007). Is there more to complicated grief than depression and posttraumatic stress disorder? A test of incremental validity. *Journal of Abnormal Psychology*, 116(2), 342 - 351.

Bowlby, J. (1980). *Loss, sadness and depression.* New York: Basic Books.

Boyle, G. J. (1991). Does item homogeneity indicate internal consistency or item redundancy in psychometric scales? *Personality and Individual Differences*, 12(3), 291 - 294.

Browne, T., Evangeli, M., & Greenberg, N. (2012). Trauma - related guilt and posttraumatic stress among journalists. *Journal of Traumatic Stress*, 25(2), 207 - 210.

Bryman, A. (2006). Integrating quantitative and qualitative research: how is it done? *Qualitative Research*, 6, 97 - 113.

Bybee, J., & Zigler, E. (1996). Guilt, guilt - evoking events, depression, and eating disorders. *Current Psychology*, 15(2), 113 - 127.

Cacciatore, J. (2010). The Unique experiences of women and their families after the death of a baby. *Social Work in Health Care*, 49(2), 134 – 148.

Cattell, R. B. (1966). The scree test for the number of factors. *Multivariate Behavioral Research*, 142, 559 – 563.

Chiu, Y. C. (2010). What drives patients to sue doctors? The role of cultural factors in the pursuit of malpractice claims in Taiwan. *Social Science and Medicine*, 71(4), 702 – 707.

Cicerelli, V. (1993). Attachment and obligation as daughters' motives for care giving behavior and subsequent effect on subjective burden. *Psychology and Aging*, 8, 144 – 155.

Collins, K. M. T., Onwuegbuzie, A. J., & Sutton, I. L. (2006). A model incorporating the rationale and purpose for conducting mixed methods research in special education and beyond. *Learning Disabilities: A Contemporary Journal*, 4, 67 – 100.

Constantino, R. E., & Bricker, P. L. (1996). Nursing postvention for spousal survivors of suicide. *Issues in Mental Health Nursing*, 17(2), 131 – 152.

Corley, M. C., Elswick, R. K., Gorman, M., & Clor, T. (2000). Development and evaluation of a moral distress scale. *Methodological Issues In Nursing Research*, 33(2), 250 – 256.

Costello, A. B., & Osborne, J. W. (2005). Best practices in Exploratory Factor Analysis: Four recommendations for getting the most from your analysis. *Practical Assessment, Research and Evaluation*, 10(7), 1 – 9.

Costello, J. (1999). "Filling the void": Grief and loneliness in older adults. *Illness, Crisis and Loss*, 7(3), 218 – 232.

Daniel, Y., A. (1985). Older and younger adults' attributions of responsibility toward rape victims and rapists. *Canadian Journal of Behavioral Science*, 17(4), 327 – 338.

Davis, C. G., Lehman, D. R., Wortman, C. B., & Silver, R. C. (1995). The undoing of traumatic life events. *Personality and Social Psychology Bulletin*, 21(2), 109 – 124.

De Groot, M., De Keijser, J., Neeleman, J., Kerkhof, A., Nolen, W., & Burger, H. (2007). Cognitive behavior therapy to prevent complicated grief among relatives and spouses bereaved by suicide: Cluster randomised controlled trial. *British Medical Journal*, 334(7601), 994 – 996.

De Vellis, R. F. (2003). *Scale development: Theory and applications* (2nd ed.). Thousand Oaks, CA: Sage.

Dewey, J. (1925). *Experience and nature.* Whitefish, MT: Kessinger.

Downey, G., Silver, R. C., & Wortman, C. B. (1990). Reconsidering the attribution – adjustment relation following a major negative event: Coping with the loss of a child. *Journal of Personality & Social Psychology*, 59(5), 925 – 936.

Drenovsky, C. K. (1994). Anger and the desire for retribution among bereaved parents. *Omega: Journal of Death and Dying*, 29(4), 303 – 312.

Dyregrov, K. (2004). Bereaved parents' experience of research participation. *Social Science and Medicine*, 58, 391 – 400.

Eatough, V., & Smith, J. A. (2006). "I was like a wild person": Understanding feelings of anger using interpretative phenomenological analysis. *British Journal of Psychology*, 97, 483 – 498.

Evans, J. R., & Mathur, A. (2005). The value of online surveys. *Internet Research*, 15(2), 195 – 219.

Exline, J. J., Park, C. L., Smyth, J. M., & Carey, M. P. (2011). Anger toward God: Social – cognitive predictors, prevalence, and links with adjustment to bereavement and cancer. *Journal of Personality and Social Psychology*, 100(1), 129 – 148.

Feilzer, Y. M. (2010). Doing mixed methods research pragmatically: Implications for the rediscovery of pragmatism as a research paradigm. *Journal of Mixed Methods Research*, 4(1), 6 – 16.

Fergus, T. A., Valentiner, D. P., McGrath, P. B., & Jencius, S. (2010). Shame – and guilt – proneness: Relationships with anxiety disorder symptoms in a clinical sample. *Journal of Anxiety Disorders*, 24(8), 811 – 815.

Field, N. P., & Bonanno, G. A. (2001). The role of blame in adaptation in the first 5 years following the death of a spouse. *American Behavioral Scientist*, 44(5), 764 – 781.

Field, N. P., Bonanno, G. A., Williams, P., & Horowitz, M. J. (2000). Appraisals of blame in adjustment in conjugal bereavement. *Cognitive Therapy and Research*, 24(5), 551 – 569.

Fischer, J. (2006). Suicide survivor guilt: The relationship between interpersonal guilt and complicated grief in suicide bereavement. *Dissertation Abstracts International: Section B: The Sciences and Engineering*, 67(2 – B), 1147.

Fleming, S., & Robinson, P. (2001). Grief and cognitive – behavioral therapy: The reconstruction of meaning. In M. S. Stroebe, R. O. Hansson, W. Stroebe & H. Schut (Eds.), *Handbook of bereavement research: Consequences, coping, and care* (pp. 647 – 670). Washington, DC: American Psychological Association.

Frazier, P. , & Tix, A. P. (2004). Testing moderator and mediator effects in counseling psychology. *Journal of Counseling Psychology*, 51(1), 115 – 134.

Freud, S. (1963). Mourning and melancholia. In P. Rieff(Ed.), P. Riviere, (Trans.). *General psychological theory* (pp. 164 – 179). New York: Macmillan. (Original work published: 1917).

Gamino, L. A. , Sewell, K. W. , & Easterling, L. W. (2000). Scott and White grief study – Phase 2: Toward an adaptive model of grief. *Death Studies*, 24(7), 633 – 660.

Gans, D. , & Silverstein, M. (2006). Norms of filial responsibility for aging parents across time and generations. *Journal of Marriage and Family*, 68, 961 – 976.

Garnefski, N. , & Kraaij, V. (2009). Cognitive coping and psychological adjustment in different types of stressful life events. *Individual Differences Research*, 7(3), 168 – 181.

Gillies, J. , & Neimeyer, R. (2006). Loss, grief, and the search for significance: Toward a model of meaning reconstruction in bereavement. *Journal of Constructivist Psychology*, 19 (1), 31 – 65.

Gojowy, A. , Kowalska, A. , Skowronek, R. , Niedzielska, H. , Ukrainska, M. , Rzepka, M. , . . . Krupka – Matuszczyk, I. (2011). Sense of guilt and anxiety in depression. *European Psychiatry*, 26, 644 – 644.

Golden, A. M. J. , & Dalgleish, T. (2012). Facets of pejorative self – processing in complicated grief. *Journal of Consulting and Clinical Psychology*, 80(3), 512 – 524.

Graham, S. , Doubleday, C. , & Guarino, P. A. (1984). The development of relations between perceived controllability and the emotions of pity, anger and guilt. *Child Development*, 55 (2), 561 – 565.

Greene, J. C. , Caracelli, V. J. , & Graham, W. F. (1989). Toward a conceptual framework for mixed – method evaluation designs. *Educational Evaluation and Policy Analysis*, 11, 255 – 274.

Guba, E. G. (1990). The alternative paradigm dialog. In E. G. Guba(Ed.), *The Paradigm Dialog.* (pp. 17 – 30). Newbury Park, CA: Sage.

Gupta, R. K. , Kulhara, P. , & Verma, S. K. (1988). Guilt and hostility in depression. *Journal of*

Personality and Clinical Studies,4(2),127 - 132.

Harder,D. H. ,Cutler,L. ,& Rockart,L. (1992). Assessment of shame and guilt and their relationships to psychopathology. *Journal of Personality Assessment*,59(3),584 - 604.

Harder,D. H. ,& Zalma,A. (1990). Two promising shame and guilt scales:A construct validity comparison. *Journal of Personality Assessment*(55),729 - 745.

Harwood,D. ,Hawton,K. ,Hope,T. ,& Jacoby,R. (2002). The grief experiences and needs of bereaved relatives and friends of older people dying through suicide:A descriptive and case - control study. *Journal of Affective Disorders*,72(2),185 - 194.

Hasui,C. ,& Kitamura,T. (2004). Aggression and guilt during mourning by parents who lost an infant. *Bulletin of the Menninger Clinic*,68(3),245 - 259.

Hazzard,A. ,Weston,J. ,& Gutterres,C. (1992). After a child's death:Factors related to parental bereavement. *Journal of Developmental and Behavioral Pediatrics*,13(1),24 - 30.

Held,P. ,Owens,G. P. ,Schumm,J. A. ,Chard,K. M. ,& Hansel,J. E. (2011). Disengagement coping as a mediator between trauma - related guilt and PTSD severity. *Journal of Traumatic Stress*,24(6),708 - 715.

Henning,K. R. ,& Frueh,B. C. (1997). Combat guilt and its relationship PTSD symptoms. *Journal of Clinical Psychology*,53(8),801 - 808.

Ho,D. Y. F. (1996). Filial piety and its psychological consequences. In M. H. Bond(Ed.),*The Handbook of Chinese Psychology*(pp. 155 - 165). Hong Kong:Oxford University Press.

Hogan,N. S. ,Greenfield,D. A. ,& Schmidt,L. A. (2001). Development and validation of the Hogan Grief Reaction Checklist. . *Death Studies*,25,1 - 32.

Horowitz,M. J. ,Weiss,D. S. ,Kaltreider,N. ,Krupnick,J. ,Marmar,C. ,Wilner,N. ,& Dewitt, K. (1984). Reactions to the death of a parents:Results from patients and field subjects. *The Journal of Nervous and Mental Disease*,172(7),383 - 392.

Hu,L. T. ,& Bentler,P. M. (1999). Cutoff criteria for fit indexes in covariance structure analysis:Conventional criteria versus new alternatives. *Structural Equation Modeling*,6(1),1 - 55.

Huang,S. T. ,& Tang,Y. Y. (1996). The loss - grief reaction and coping behavior of a mother whose child died of a brain tumor. *The Journal of Nursing(In Chinese)*,43(4),34 - 41.

Hwang,K. K. (1987). Face and favor:The Chinese power game. *The American Journal of Soci-*

ology, 92(4), 944 - 974.

Hwang, K. K. (2001). Morality: East and West. In N. J. Smelser & P. B. Baltes (Eds.), *International encyclopedia of the social and behavior science* (pp. 10039 - 10043). Amsterdam: Elsevier.

Hyrkäs, K., Kaunonen, M., & Paunonen, M. (1997). Recovering from the death of a spouse. *Journal of Advanced Nursing*, 25(4), 775 - 779.

Irion, J. C., & Balanchard - Rields, F. (1987). A cross - sectional comparison of adaptive coping in adulthood. *Journal of Gerontology*, 42(5), 502 - 504.

Janoff - Bulman, R. (1979). Characterological versus behavioral self - blame: Inquiries into depression and rape. *Journal of Personality and Social Psychology*, 37, 1798 - 1809.

Janoff - Bulman, R. (1992). *Shattered assumptions: Towards a new psychology of trauma.* New York: The Free Press.

Jarrett, R. B., & Weissenburger, J. E. (1990). Guilt in depressed outpatients. *Journal of Consulting and Clinical Psychology*, 58(4), 495 - 498.

Johnson, R. B., & Onwuegbuzie, A. J. (2004). Mixed methods research: A research Paradigm whose time has come. *Educational Researcher*, 33(7), 14 - 26.

Johnson, R. B., Onwuegbuzie, A. J., & Turner, L. A. (2007). Toward a definition of mixed methods research. *Journal of Mixed Methods Research*, 1(2), 112 - 133.

Johnson, R. C., Danko, G. P., Huang, Y., Park, J. Y., Jonhson, S. B., & Nagoshi, C. T. (1987). Guilt, shame and adjustment in three cultures. *Personality and Individual Differences*, 8, 357 - 364.

Jonasson, J. M., Hauksdottir, A., Nemes, S., Surkan, P. J., Valdimarsdottir, U., Onelov, E., & Steineck, G. (2011). Couples' communication before the wife's death to cancer and the widower's feelings of guilt or regret after the loss - A population - based investigation. *European Journal of Cancer*, 47, 1564 - 1570.

Jordan, J. R. (2001). Is suicide bereavement different? A reassessment of the literature. *Suicide and Life Threatening Behavior*, 31(1), 91 - 102.

Kenny, D. A., Kashy, D. A., & Bolger, N. (1998). Data analysis in social psychology. In D. T. Gilbert, S. T. Fiske & G. Lindzey (Eds.), *The handbook of social psychology* (4th ed., pp.

233 – 265). New York: Oxford University Press.

Kerr, R. B. (1994). Meanings to adult daughters attach to a parent's death. *Western Journal of Nursing Research*, 16(4), 347 – 365.

Kersting, A., Brähler, E., Glaesmer, H., & Wagner, B. (2011). Prevalence of complicated grief in a representative population – based sample. *Journal of Affective Disorders*, 131(1 – 3), 339 – 343.

Kim, S. H. (2009). The influence of finding meaning and worldview of accepting death on anger among bereaved older spouses. *Aging and Mental Health*, 13(1), 38 – 45.

Kim, U. (1994). Individualism and collectivism: Conceptual clarification and elaboration. In U. Kim, H. C. Triandis, C. Kagitcibasi & G. Yoon (Eds.), *Individualism and collectivism* (pp. 19 – 49). California: Thousand Oaks.

Klass, D., & Chow, A. Y. M. (2011). Culture and ethnicity in experiencing, policing and handling grief. In R. A. Neimeyer, D. L. Harris, H. R. Winokuer & G. F. Thornton (Eds.), *Grief and bereavement in contemporary society* (pp. 341 – 354). New York: Rouledge.

Kline, R. B. (2005). *Principles and practice of structural equation modeling* (2nd ed.). New York: Guilford Press.

Kochanska, G., Gross, J. N., Lin, M., & Nichols, K. E. (2002). Guilt in young children: Development, determinants, and relations with a broader system of standards. *Child Development*, 73, 461 – 482.

Kolb, L. (1983). Return of the repressed: Delayed stress reaction to war. *Journal of American Academy of Psychoanalysis*, 11, 531 – 545.

Kowalski, S. D., & Bondmass, M. D. (2008). Physiological and psychological symptoms of grief in widows. *Research in Nursing and Health*, 31(1), 23 – 30.

Kristensen, P., Weisæth, L., & Heir, T. (2010). Predictors of complicated grief after a natural disaster: A population study two years after the 2004 south – east Asian tsunami. *Death Studies*, 34(2), 137 – 150.

Kubany, E. S. (2003). Guilt: Elaboration of a multidimensional model. *The Psychological Record*, 53(1), 51 – 90.

Kubany, E. S., & Abueg, F. R. (1996). Development and validation of the Trauma – Related Guilt Inventory (TRGI). *Psychological Assessment*, 8(4), 428 – 444.

Kubany, E. S., Abueg, F. R., Kilauano, W. L., Manke, F. P., & Kaplan, A. S. (1997). Development and validation of the sources of Trauma – Related Guilt Survey – war – zone version. *Journal of Traumatic Stress*, 10(2), 235 – 258.

Kubany, E. S., & Manke, F. P. (1995). Cognitive therapy for trauma – related guilt: Conceptual bases and treatment outlines. *Cognitive and Behavioral Practice*, 2(1), 27 – 61.

Kugler, K., & Jones, W. H. (1992). On conceptualizing and assessing guilt. *Journal of Personality and Social Psychology* 62(2), 318 – 327.

Lang, A., Gottlieb, L. N., & Amsel, R. (1996). Predictors of husbands" and wives" grief reactions following infant death: The role of marital intimacy. *Death Studies*, 20(1), 33 – 57.

Larkin, M., Watts, S., & Clifton, E. (2006). Giving voice and making sense in interpretative phenomenological analysis. *Qualitative Research in Psychology*, 3, 102 – 120.

Leary, M. R. (2007). How the self became involved in affective experience: Three sources of self – reflective emotions. In J. L. Tracy, R. W. Robins & J. P. Tangney (Eds.), *The self – conscious emotions* (pp. 38 – 52). New York: The Guilford Press.

Lehman, D. R., & Wortman, C. B. (1987). Long – term effects of losing a spouse or child in a motor vehicle crash. *Journal of Personality and Social Psychology*, 52(1), 218 – 231.

Leung, C. M., Ho, S., Kan, C. S., Hung, C. H., & Chen, C. N. (1993). Evaluation of the Chinese version of the Hospital Anxiety and Depression Scale. *International Journal of Psychosomatics*, 40, 29 – 34.

Li, J., Chan, C. L. W., & Chow, A. Y. M. (2011, November). *The grief and guilt experience among bereaved college students.* Paper presented at the the 18th Hong Kong International Cancer Congress, Hong Kong.

Li, J., Wang, L., & Fischer, K. W. (2004). The organization of Chinese shame concepts. *Cognition and Emotion*, 18(6), 767 – 797.

Lincoln, Y. S., & Guba, E. G. (2000). Three epistemological stances for qualitative inquiry. In N. Denzin & Y. S. Lincoln (Eds.), *Handbook of Qualitative Research* (2nd ed., pp. 189 – 213). London: Sage Publications, Inc.

Lindemann, E. (1944). Symptomatology and management of acute grief. *American Journal of Psychiatry*, 101, 141 – 148.

Lobb, E. A. , Kristjanson, L. J. , Aoun, S. M. , Monterosso, L. , Halkett, G. K. B. , & Davies, A. (2010). Predictors of complicated grief: A systematic review of empirical studies. *Death Studies*, 34(8), 673 – 698.

Losada, A. , Márquez – González, M. , Peñacoba, C. , & Romero – Moreno, R. (2010). Development and validation of the Caregivers Guilt Questionnaire. *International Psychogeriatrics*, 22 (4), 650 – 660.

Lowe, C. M. (1964). The equivalence of guilt and anxiety as psychological constructs. *Journal of Consulting Psychology*, 28(6), 553 – 554.

Luby, J. , Belden, A. , Sullivan, J. , Hayen, R. , McCadney, A. , & Spitznagel, E. (2009). Shame and guilt in preschool depression: evidence for elevations in self – conscious emotions in depression as early as age 3. *Journal of Child Psychology and Psychiatry*, 50(9), 1156 – 1166.

Mallimson, K. E. (2006). Survivor guilt and posttraumatic stress disorder 18 months after September 11, 2001: Influences of prior trauma, exposure to the event, and bereavement. *Dissertation Abstracts International: Section B: The Sciences and Engineering*, 66(9 – B), 5096.

Marks, N. F. , Jun, H. , & Jieun, S. (2005). *Death of parents and adult psychological and physical health: A prospective national study*. working paper. Center for Demography and Ecology, University of Wisconsin – Madison.

Markus, H. R. , & Kitayama, S. (1991). Culture and the self: Implications for cognition, emotion, and motivation. *Psychological Review*, 98(2), 224 – 253.

Marsh, H. W. , Balla, J. R. , & McDonald, R. P. (1988). Goodness of fit indexes in confirmatory factor analysis: the effect of sample size. *Psychological Bulletin*, 103, 391 – 410.

Martin, M. , Sadlo, G. , & Stew, G. (2006). The phenomenon of boredom. *Qualitative Research in Psychology*, 3, 193 – 211.

Massey, S. , Cameron, A. , Ouellette, S. , & Fine, M. (1998). Qualitative approaches to the study of thriving: What can be learned?. *Journal of Social Issues*, 54, 337 – 355.

McMenamy, J. M. , Jordan, J. R. , & Mitchell, A. M. (2008). What do suicide survivors tell us they need? Results of a pilot study. *Suicide and Life – Threatening Behavior*, 38(4), 375 – 389.

Melhem, N. M. , Day, N. , Shea, M. K. , Day, R. , Reynolds Iii, C. F. , & Brent, D. (2004). Predictors of complicated greif among adolescents exposed to a peer's suicide. *Journal of Loss and Trauma*,9(1),21 - 34.

Miles, M. S., & Demi, A. S. (1983). Toward the development of a theory of bereavement guilt: Sources of guilt in bereaved parents. *Omega: Journal of Death and Dying*, 14(4), 299 - 314.

Miles, M. S., & Demi, A. S. (1991—1992). A comparison of guilt in bereaved parents whose children died by suicide, accident, or chronic disease. *Omega: Journal of Death and Dying*, 24(3), 203 - 215.

Mizota, Y., Ozawa, M., Yamazaki, Y., & Inoue, Y. (2006). Psychosocial problems of bereaved families of HIV - infected hemophiliacs in Japan. *Social Science and Medicine*, 62(10), 2397 - 2410.

Morgan, D. L. (2007). Paradigms lost and pragmatism regained: Methodological implications of combining qualitative and quantitative methods. *Journal of Mixed Methods Research*, 1(1), 48 - 76.

Murphy, H. B. (1978). The advent of guilt feelings as a common depressive symptom: A historical comparison on two continents. *Psychiatry*, 41(3), 229 - 242.

Neimeyer, R., Harris, D. L., Winokuer, H. R., & Thornton, G. F. (2011). *Grief and bereavement in contemporary society*. New York: Routledge.

Neimeyer, R., & Hogan, N. S. (2001). Quantitative or qualitative? Measurement issues in the study of grief. In M. S. Stroebe, R. O. Hansson, W. Stroebe & H. Schut (Eds.), *Handbook of bereavement research: Consequences, coping and care* (pp. 89 - 118). Washington, DC: American Psychological Association.

Newman, I., Ridenour, C. S., Newman, C., & Jr. GeMarco, G. M. P. (2003). A typology of research purposes and its relationship to mixed methods. In A. Tashakkori & C. Teddlie (Eds.), *Handbook of mixed methods in social and behavioral research* (pp. 167 - 188). London: Sage Publications.

Newson, R. S., Boelen, P. A., Hek, K., Hofman, A., & Tiemeier, H. (2011). The prevalence and characteristics of complicated grief in older adults. *Journal of Affective Disorders*, 132(1 - 2), 231 - 238.

Nixon, G. F., & Steffeck, J. (1976). Relationship between anxiety and guilt. *Psychological Reports*, 38(3), 766 - 766.

O'Connor, L. E., Berry, J. W., Weiss, J., Bush, M., & Sampson, H. (1997). Interpersonal guilt: The development of a new measure. *Journal of Clinical Psychology*, 53(1), 73 - 89.

Onwuegbuzie, A. J., Bustamante, R. M., & Nelson, J. A. (2010). Mixed research as a tool for developing quantitative instruments. *Journal of mixed methods research*, 4(1), 56 - 78.

Opp, R. E., & Samson, A. Y. (1989). Taxonomy of guilt for combat veterans. *Professional Psychology: Research and Practice*, 20(3), 159 - 165.

Orth, U., Robins, R. W., & Soto, C. J. (2010). Tracking the trajectory of shame, guilt, and pride across the life span. *Journal of Personality and Social Psychology*, 99(6), 1061 - 1071.

Ott, C. H. (2003). The impact of complicated grief on metal and physical health at various points in the bereavement process. *Death Studies*, 27(3), 249 - 272.

Owens, G. P., Steger, M. F., Whitesell, A. A., & Herrera, C. J. (2009). Posttraumatic stress disorder, guilt, depression, and meaning in life among military veterans. *Journal of Traumatic Stress*, 22(6), 654 - 657.

Parkes, C. M. (1996). *Bereavement: Studies of grief in adult life*. London: Routledge.

Parkes, C. M. (2008). Bereavement following disasters. In M. S. Stroebe, R. O. Hansson, H. Schut & W. Stroebe (Eds.), *Handbook of bereavement research and practice: Advances in theory and practice* (pp. 463 - 484). Washington, DC: American Psychological Association.

Patton, M. Q. (1990). *Qualitative evaluation and research methods*. Newbury Park, CA: Sage Publications.

Pereda, N., Arch, M., Peró, M., Guàrdia, J., & Forns, M. (2011). Assessing guilt after traumatic events: The Spanish adaptation of the Trauma - Related Guilt Inventory. *European Journal of Psychological Assessment*, 27(4), 251 - 257.

Piaget, J. (1932). *The moral judgment of the child.* New York: Harcourt, Brace.

Prigerson, H. G., & Frank, E. (1995). Complicated grief and bereavement - related depression as distinct disorders: Preliminary empirical validation in elderly bereaved spouses. *American Journal of Psychiatry*, 152(1), 22 - 30.

Prigerson, H. G., Maciejewski, P. K., Reynolds, C. F., Bierhals, A. J., Newsom, J. T., Fasiczka, A., ... Miller, M. (1995). Inventory of complicated grief: A scale to measure maladaptive symptoms of loss. *Psychiatry Research*, 59(1 - 2), 65 - 79.

Prigerson, H. G., Shear, M. K., Jacob, G. A., Reynolds, C. F., Maciejewski, P. K., Davidson,

J. R. ,…Zisook,S. (1999). Consenseus criteria for traumatic grief. *British Journal of Psychiatry*,174,67 – 73.

Prigerson,H. G. ,Vanderwerker,L. C. ,& Maciejewski,P. K. (2008). A case for inclusion of prolonged grief disorder in DSM – V. In M. S. Stroebe,R. O. Hansson,H. Schut & W. Stroebe (Eds.),*Handbook of bereavement research and practice:Advances in theory and practice*(pp. 165 – 186). Washington,DC:American Psychological Association.

Rando,T. A. (1993). *Treatment of complicated mourning*. Champaign,Illinois:Research Press.

Rorty,R. (1999). *Philosophy and social hope.* London:Penguin Books.

Rosenberg,M. (1965). *Society and the adolescent self – image.* Princeton,NJ:Princeton University Press.

Rosenblatt,P. C. (2008). Grief across cultures:A review and research agenda. In M. S. Stroebe,R. O. Hansson,H. Schut & W. Stroebe(Eds.),*Handbook of bereavement research and practice:Advances in theory and intervention*(pp. 207 – 222). Washington, DC: American Psychological Association.

Rothbaum,F. ,& Xu,X. (1995). The theme of giving back to parents in Chinese and American songs. *Journal of Cross – Cultural Psychology*,26,698 – 713.

Russell. (1991). Culture and the categorization of emotions. *Psychological Bulletin*,110(3), 426 – 450.

Sale,J. E. M. ,Lohfeld,L. H. ,& Brazil,K. (2002). Revisiting the quantitative – qualitative debate:Implications for mixed – methods research. *Qunality and Quantity*,36,43 – 53.

Sanders,C. M. ,Mauger,P. A. ,& Strong,P. N. (1985). *A manual for the Grief Experience Inventory.* Palo Alto,CA:Consulting Psychologists Press.

Scharlach,A. E. (1991). Factors associated with filial grief following the death of an elderly patient. *American Journal of Orthopsychiatry*,61,307 – 313.

Schwab,R. (1996). Gender differences in parental grief. *Death Studies*,20(2),103 – 113.

Shanfield,S. B. ,& Swain,B. J. (1984). Death of adult children in traffic accidents. *Journal of Nervous and Mental Disease*,172(9),533 – 538.

Shapiro,L. J. ,& Stewart,S. E. (2011). Pathological guilt:A persistent yet overlooked treatment factor in obsessive – compulsive disorder. *Annals of Clinical Psychiatry*,23(1),63 – 70.

Shaver, P. R., Wu, S., & Schwartz, J. C. (1992). Cross – cultural similarities and differences in emotion and its representation: A prototype approach. In M. S. Clark (Ed.), *Review of personality and social psychology* (Vol. 13, pp. 175 – 212). Newbury Park, CA: Sage.

Shear, K., Monk, T., Houck, P., Melhem, N., Frank, E., Reynolds, C., & Sillowash, R. (2007). An attachment – based model of complicated grief including the role of avoidance. *European Archives of Psychiatry and Clinical Neuroscience*, 257(8), 453 – 461.

Shear, M. K., Simon, N., Wall, M., Zisook, S., Neimeyer, R., Duan, N., ··· First, M. (2011). Complicated grief and related bereavement issues for DSM – 5. *Depression and Anxiety* 28 (2), 103 – 117.

Smith, J. A. (1996). Beyond the divide between cognition and discourse: using interpretative phenomenological analysis in health psychology. *Psychology and Health*, 11(2), 261 – 271.

Smith, J. A., Larkin, M., & Flowers, P. (2009). *Interpretative phenomenological analysis: Theory, method and research: understanding method and application.* London: Sage.

Smith, M. E., Nunley, B. L., Kerr, P. L., & Galligan, H. (2011). Elders' experiences of the death of an adult child. *Issues in Mental Health Nursing*, 32(9), 568 – 574.

Stewart, M., Craig, D., MacPherson, K., & Alexander, S. (2001). Promoting positive affect and diminishing loneliness of widowed seniors through a support intervention. *Public Health Nursing*, 18(1), 54 – 63.

Strauss, A., & Corbin, J. (1998). *Basics of qualitative research: Techniques and procedures for developing grounded theory.* London: SAGE Publications.

Street, A. E., Gibson, L. E., & Holohan, D. R. (2005). Impact of childhood traumatic events, trauma – related guilt, and avoidant coping strategies on PTSD symptoms in female survivors of domestic violence. *Journal of Traumatic Stress*, 18(3), 245 – 252.

Strelan, P. (2007). Who forgives others, themselves, and situations? The roles of narcissism, guilt, self – esteem, and agreeableness. *Personality and Individual Differences*, 42(2), 259 – 269.

Stroebe, M. S., Folkman, S., Hansson, R. O., & Schut, H. (2006). The prediction of bereavement outcomes: Development of an integrative risk factor framework. *Social Science and Medicine*, 63, 2440 – 2451.

Stroebe, M. S., Hansson, R., Stroebe, W., & Schut, H. (2001). Introduction: Concepts and issues in contemporary research on bereavement. In M. Stroebe, R. Hansson, W. Stroebe & S. Henk (Eds.), *Handbook of bereavement research: Consequences, coping, and care* (pp. 3 – 22). Washington, DC: American Psychological Association.

Stroebe, M. S., Hansson, R. O., Schut, H., & Stroebe, W. (2008). *Handbook of bereavement research and practice: Advances in theory and practice.* Washington DC: American Psychological Association.

Stroebe, W., & Schut, H. (2001). Risk factors in bereavement outcome: A methodological and empirical review. In M. S. Stroebe, R. O. Hansson, S. Wolfgang & S. Henk (Eds.), *Handbook of bereavement research: Consequences, coping and care* (pp. 349 – 371). Washington, DC: American Psychological Association.

Stroebe, W., Stroebe, M. S., Abakoumkin, G., & Schut, H. (1996). The role of loneliness and social support in adjustment to loss: A test of attachment versus stress theory. *Journal of Personality & Social Psychology*, 70(6), 1241 – 1249.

Surkan, P. J., Kreicbergs, U., Valdimarsdóttir, U., Nyberg, U., Onel v, E., Dickman, P. W., & Steineck, G. (2006). Perceptions of inadequate health care and feelings of guilt in parents after the death of a child to a malignancy: A population – based long – term follow – up. *Journal of Palliative Medicine*, 9(2), 317 – 331.

Tangney, J. P. (1990). Assessing individual differences in proneness to shame and guilt: Development of the Self – Conscious Affect and Attribution Inventory. *Journal of Personality and Social Psychology*, 59(1), 102 – 111.

Tangney, J. P. (1999). The Self – conscious emotions: Shames, guilt, embarrassment and pride. In T. Dalgleish & M. J. Power (Eds.), *Handbook of Cognition and Emotion* (pp. 541 – 568). Chichester, England: John Wiley & Sons.

Tangney, J. P., Burggraf, S. A., & Wanger, P. E. (1995). Shame – proneness, guilt – proneness, and psychological symptoms. In J. P. Tangney & K. W. Fisher (Eds.), *Self – conscious emotions: Shame, guilt, embarrassment, and pride.* (pp. 343 – 367). New York: Guilford.

Tangney, J. P., & Dearing, R. L. (2002). *Shame and guilt.* New York: Guilford Press.

Tangney, J. P., & Fischer, K. W. (1995). *Self – conscious emotions: The psychology of shame, guilt, embarrassment, and pride.* New York: Guilford Press.

Tangney, J. P. , Stuewig, J. , & Mashek, D. J. (2007). Moral emotions and moral behavior. *Annual Review of Psychology*, 58, 345 – 372.

Teddlie, C. , & Tashakkori, A. (2003). Major issues and controversies in the use of mixed methods in the social and behavioral sciences. In A. Tashakkori & C. Teddlie (Eds.), *Handbook of mixed methods in social and behavioral research* (pp. ix – xv). London: Sage

Tilghman – Osborne, C. , Cole, D. A. , & Felton, J. W. (2010). Definition and measurement of guilt: Implications for clinical research and practice. *Clinical Psychology Review*, 30(5), 536 – 546.

Tolstikova, K. (2009 – 2010). Internet method in bereavement research: Comparison of online and offline surveys. *Omega: Journal of Death and Dying*, 60(4), 327 – 349.

Tooley, K. (1975). The choice of a surviving sibling as "scapegoat" in some cases of maternal bereavement: A case report. *Journal of Child Psychology and Psychiatry and Allied Disciplines*, 16(4), 331 – 339.

Torges, C. M. , Stewart, A. J. , & Nolen – Hoeksema, S. (2008). Regret resolution, aging, and adapting to loss. *Psychology and Aging*, 23(1), 169 – 180.

Tracy, J. L. , & Robins, R. W. (2007). The self in Self – conscious emotions: A cognitive appraisal approach. In J. L. Tracy, R. W. Robins & J. P. Tangney (Eds.), *The Self – Conscious Emotion* (pp. 3 – 20). New York: The Guilford Press.

Umberson, D. (1992). Relationships between adult children and their parents: Psychological consequences for both generations. *Journal of Marriage and the Family*, 54(664 – 674).

Umberson, D. (2003). *Death of a parents: Transition to a new adult identity*. Cambridge: Cambridge University Press.

Umberson, D. , & Chen, M. (1994). Effects of a parents' death on adult children: Relationship salience and reaction to loss. *American Sociological Review*, 59(1), 152 – 168.

van der Houwen, K. , Stroebe, M. S. , Stroebe, W. , Schut, H. , van den Bout, J. , & Meij, L. W. – D. (2010). Risk factors for bereavement outcome: A multivariate approach. *Death Studies*, 34(3), 195 – 220.

Wang, L. L. , Fan, X. , & Wilson, V. L. (1996). Effects of nonnormal data on parameter estimates for a model with latent and manifest variables: An empirical study. *Structural Equation*

Modeling,3(3),228 – 247.

Weinberg, N. (1994). Self – blame, other blame, and desire for revenge: Factors in recovery from bereavement. *Death Studies*,18(6),583 – 593.

Weinberg, N. (1995). Does apologizing help? The role of self – blame and making amends in recovery from bereavement. *Health and Social Work*,20(4),294 – 294.

Wijngaards – de Meij, L., Stroebe, M. S., Schut, H., Stroebe, W., van den Bout, J., van der Heijden, P., & Dijkstra, I. (2005). Couples at risk following the death of their child: Predictors of grief versus depression. *Journal of Consulting and Clinical Psychology*,73,617 – 623.

Willig, C. (2001). *Introducing qualitative research in psychology: Adventures in theory and method*. Philadelphia, PA: Open University Press.

Winsborough, H. H., Larry, L. B., & William, S. A. (1991). *The death of parents and the transition to old age*. NSFH – 39. working paper: NSFH – 39. University of Wisconsin Center for Demography and Ecology. Madison, WI.

Wohl, M. J., & DeShea, L. (2008). Looking within: Measuring state self – forgiveness and its relationship to psychological well – being. *Canadian Journal of Behavioral Science*,40(1),1 – 10.

Wong, O. M. H., & Chau, B. H. P. (2006). The evolving role of filial piety in eldercare in Hong Kong. *Asian Journal of Social Psychology*,34(4),600 – 617.

Woo, I. M. H., & Chan, C. L. W. (2010). Management of survival guilt by a Chinese widower through the use of continuing bonds. *Mortality*,15(1),38 – 46.

Yang, K. S., Yeh, K. – H., & Huang, L. L. (1989). A social attitudinal analysis of Chinese filial piety: Conceptualization and assessment. *Bulletin of the Institute of Ethnology*,56,171 – 227(in Chinese).

Yeh, K. H. (2002). The beneficial and harmful effects of filial piety: An integrative analysis. In K. S. Yang, K. K. Hwang, P. B. Pedersen & I. Daibo(Eds.), *Progress in asian social psychology*(pp. 67 – 82). London: Westport, Connecticut.

Yeh, K. H., & Bedford, O. (2003). A test of the dual filial piety model. *Asian Journal of Social Psychology*,6,215 – 228.

Ylitalo, N., Valdimarsdottir, U., Onelov, E., Dickman, P. W., & Steineck, G. (2008). Guilt after the loss of a husband to cancer: Is there a relation with the health care provided? *Acta*

Oncologica,47(5),870 – 878.

Zechmeister, J. S., & Romero, C. (2002). Victim and offender accounts of interpersonal conflict: Autobiographical narratives of forgiveness and unforgiveness. *Journal of Personality and Social Psychology*,82(4),675 – 686.

Zimmerman, J. (2001). Dyspepsia as a somatic expression of guilt: A case report. *American Journal of Clinical Hypnosis*,44(1),57 – 61.

Zisook, S., & Shuchter, S. R. (1985). Time course of spousal bereavement. *General Hospital Psychiatry*,7(2),95 – 100.

[illegible]

[illegible] (1987). [illegible]

[illegible] (2002). [illegible] consequences of [illegible]

[illegible]

[illegible] Social Psychology, [illegible]

[illegible]

[illegible] Journal of [illegible]

[illegible]

[illegible]